Consider a Spherical Cow

*A Course
in Environmental
Problem Solving*

John Harte
University of California, Berkeley

University Science Books   Mill Valley, California

Printed in the United States of America

10 9 8 7 6 5

Library of Congress Cataloging in Publication Data

Harte, John, 1939 –
   Consider a spherical cow—

   Bibliography: p.
   Includes index.
   1. Ecology—Mathematics.   2. Ecology—Problems, exercises, etc.   I. Title.
QH541.15.M34H37   1985   304.2'01'08   85-203
ISBN 0-935702-58-X (pbk)

University Science Books
20 Edgehill Road
Mill Valley, CA 94941

To my parents, Ethel and Jerry—
with
love and admiration

It is the mark of an instructed mind to rest satisfied with the degree of precision which the nature of the subject permits and not to seek an exactness where only an approximation of the truth is possible.

—Aristotle

Contents

Preface

This book should provide a novel, and I hope enjoyable, way of learning how to use relatively simple mathematical methods (often of the "back-of-the-envelope" variety) to understand how planet Earth and its inhabitants interact.

The idea for this text evolved from courses in environmental science I have taught at Yale University and the University of California at Berkeley over the past 15 years. These courses have ranged from the introductory undergraduate to the advanced graduate level. Regardless of the level, I have stressed quantitative problem solving in all my courses, and over the past 15 years I have invented a sizeable repertoire of homework problems. These, along with a few delightful ones contributed by my colleagues, form the basis for this book.

One thing my courses have taught *me* (or rather have recalled to mind, for I knew it all too well when I was a student) is that the ordinary combination of university math and science courses does not prepare students well for solving a frequently occurring type of "word problem." Such problems call for a quantitative answer, but their solution involves information from several disciplines scrambled together. Often the statement of these problems has that fuzzy quality characteristic of real-world situations. The problems we confront outside the classroom rarely take the streamlined form most textbooks rely upon to test how well we've read each chapter.

In this book I try to provide guidance in overcoming barriers to problem solving. To that end, the problems and solutions in the first two chapters progress through a series of insights into problem solving itself. A more customary organization by environmental topic is used in the final chapter.

At the core of the book are 44 problems, with as many worked-out solutions. Chapter I provides a set of warm-up exercises. Elementary quantitative skills, such as conversion of units and approximation methods, suffice to solve these first problems. Chapter II introduces a variety of ''back-of-the-envelope'' problem-solving methods. These techniques enable you to solve many problems with very little effort—often in a few lines—once you know how to begin and what tools to use. In Chapter III, Beyond the Back of the Envelope, the problem-solving methods developed in the preceding chapters are applied to more complex situations. Methods of problem solving more advanced than those in Chapter II are introduced there. In the problems of Chapter III you will recognize the real-world fuzzy quality referred to above. To solve these we will have to define our variables and system boundaries, invent models, and select tools for extracting information from the models.

Homework exercises are provided at the end of each problem. Particularly difficult exercises are indicated by an asterisk. A few of the tasks proposed are very difficult, requiring term-paper effort; these are marked with two asterisks. Numerical answers to many of the exercises are given at the back of the book.

The solutions to certain of the 44 problems, particularly the more complex ones in Chapter III, are presented at three levels. First, I provide a ''hand-waving'' solution (i.e., an informed guess) in which the qualitative behavior of the problem's elements is deduced. Often the sign (for example, heating versus cooling) and the order of magnitude of the response of a complex system to a disturbance can be figured out without a detailed mathematical analysis. In some cases the ''hand-waving'' approach gives an absurd answer. By identifying the reason for that absurdity, we can then gain insight into how the problem should really be solved. At the second level, analytical procedures and a detailed quantitative solution are presented. Because realistic environmental issues are dealt with in this book, a number of simplifying assumptions are generally made at this level to obtain a precise solution. At the third level of solution, I describe methods for deducing the approximate consequences of removing some of level two's simplifying assumptions. Obtaining the results of such deductions is often left as a homework exercise.

I believe that high-school level mathematics, properly applied, can go a long way toward elucidating complex situations. Readers with limited or no prior exposure to calculus and differential equations will be able to follow completely most of the problem solutions presented here. Only the last two problems in Chapter II and a few in Chapter III require some use of simple differential equations; even in those problems the qualitative discussion will be of value to readers with relatively little preparation in mathematics.

Solution to some of the problems presented here requires fundamental information to which not all readers have access. The Appendix to this volume, containing tabulated data about nature and technology, should help. In addition, useful source materials are cited throughout the text for readers seeking further information. A bibliography at the back of the book lists these sources. A glossary is also provided.

All the problems presented here can be solved analytically, without the use of computers. Even in the more difficult problems, the mathematical models employed are not elaborate. I believe it is preferable in environmental analysis to develop relatively simple, analytically tractable models, rather than complex ones requiring truckloads of parameters. The advantage of being able to "tinker" mentally with a simple, penetrable model, and thus explore the consequences of a variety of assumptions, outweighs in most cases the greater realism that might be attained with a complex model.

Thus the "spherical cow" in the title of this book. The phrase comes from a joke about theoreticians I first heard as a graduate student. Milk production at a dairy farm was low so the farmer wrote to the local university, asking help from academia. A multidisciplinary team of professors was assembled, headed by a theoretical physicist, and two weeks of intensive on-site investigation took place. The scholars then returned to the university, notebooks crammed with data, where the task of writing the report was left to the team leader. Shortly thereafter the farmer received the write-up, and opened it to read on the first line: "Consider a spherical cow. . . ."

The spherical cow approach to problem solving involves the stripping away of unnecessary detail, so that only essentials remain. Of course, approaching the complex world from the spherical cow perspective can sometimes annoy others. To an expert who has labored long in the field, the cow that to you is spherical may be sacred. The trick is to know which details can be stripped away without changing the essentials. This book should help readers develop a knack for doing this.

This text should serve two functions. First, it should teach the reader how to transform realistic, qualitatively described problems into quantifiably solvable form and to arrive at an approximate solution. Second, it should teach concepts in environmental science from the novel perspective of problem solving. Readers may want to supplement this book with a general textbook in environmental science, such as the excellent *Ecoscience* by Ehrlich, Ehrlich, and Holdren (1977). Others, with more advanced backgrounds, are urged to graze among the more specialized sources cited throughout this text.

Acknowledgments

The "spherical cow" approach to problem solving is as old as science, but by reinforcing my enthusiasm for it with their own, a few teachers and colleagues contributed to the development of this book. They include Edward Purcell, Victor Weisskopf, Robert Socolow, Charles Walker, Robert Wheeler, Robert May, and John Holdren.

Early versions of the manuscript were read by Harrison Brown, Andrew Cohen, Paul Ehrlich, Steven Fetter, Peter Gleick, Ethel and Mary Ellen Harte, John Holdren, Kersten Johnson, Michael Lazarus, Michael Maniates, Robert Mann, Robert Socolow, and Kenneth Watt. I am grateful to them for their valuable comments and suggestions.

Ideas for problems and homework exercises were stimulated by conversations with Robert Adair, Andrew Gunther, Paul Ehrlich, John Holdren, Douglas McLaren, Richard Miller, James Morgan, Richard Schneider, Stephen Schneider, Robert Socolow, Frank Starr, Charles Walker, James C. G. Walker, and Thomas Powell. Others whose writings provided inspiration for problems are acknowledged in the text.

Over the years numerous students in my courses have generously shared with me their frustration and anger at the devils herein with which I plagued them. This feedback has been of great value in helping me get the "bugs" out of the problems.

For the unearthing of data I am grateful to Mari Wilson, Pi Chao Chen, John Holdren, and Anthony Nero. John Cairns supplied the apt epigraph from Aristotle. A conversation with Donald Goldsmith clinched the title. William Kaufmann provided the quotation in Chapter II from Herbert Simon and essential advice on the structure of the book. With Maggie Duncan at the helm, the manuscript weathered the storms of production in fine style. Ike Burke did a superb, insightful editing job.

I wrote most of the text during two summers at the Rocky Mountain Biological Laboratory, where my colleagues and the incomparable mountains provided a most enjoyable work place. Financial support from the Hewlett Foundation during that time is gratefully acknowledged.

Finally, I am deeply thankful to my wife, Mary Ellen, for her help in all stages of the project. She drew the original illustrations, did all the word processing of the text, located information, and thoroughly edited the entire manuscript both for grammar and content. For all of this, and for her love especially, I am lucky.

Chapter I
Warm-up Exercises

Only basic problem-solving skills are needed in this chapter. The first problem requires you to guess the approximate values of some numbers that you probably haven't thought about very much. (Guessing makes some people uncomfortable, but give it a try.) Problems I.2 and I.3 can be solved using only basic ideas about areas, volumes, and density; the second homework exercise in Problem I.3 will get you thinking about probabilities. The next two problems are about depletion and growth, determining how long resources will last at present consumption rates and how rapidly population density increases at the present growth rate. If your logarithms are rusty, now is the time to polish them up. The last two problems in this chapter show how to derive measures of the magnitude of the human presence on Earth today. Solving them requires skill in selecting relevant data and in converting from one kind of unit to another. All but the first problem in this chapter will send you scurrying to the tables of information in the Appendix. Familiarize yourself with what the Appendix offers. You'll refer to it often in solving problems later on.

1. Counting Cobblers

How many cobblers are there in the United States?[1]

. . . .

One excuse for including this problem in a book about the environment is that getting your shoes repaired consumes less resources than buying a new pair. It is here mainly, however, to illustrate the ease with which a few plausible guesses can be combined to answer a question that at first glance seems resistant to guesswork. Can you estimate the order of magnitude[2] of the answer?

To do so, you could find out if there are cobbler licensing boards and, if so, write to them for their statistics. Or you could walk to the library and check the yellow pages of telephone directories for representative U. S. cities. However, why not be lazy and let your mind do the walking? Start by assuming that cobblers are generally busy most of the work week. As a rough estimate, they spend about 10 minutes on a heel job and perhaps 30 minutes on full heels and soles. More complicated repairs are rare, so ignore them. If time out for cleaning shop and dealing with customers is included, an average of 30 minutes per job is a reasonable guess. (Remember, the answer is an order of magnitude.)

10,031, 10,032 . . .

---

1. A variation on this problem is found in the text *Used Math* by Swartz (1973), which is an excellent introductory review of applied mathematics.

2. When a number characterizing something is known imprecisely, either because the measurements are poor or because the "something" varies a lot, an order-of-magnitude estimate is often given. Usually, orders of magnitude are expressed as powers of 10; a number that is in the range of 0.3 to 3 is said to be on the order of magnitude of 1; a number between 3 and 30 is on the order of magnitude of 10, and so forth.

By this reasoning, a cobbler can finish perhaps 15 jobs in a work-day, or about 4000 a year. All you need to know now is how many repair jobs are done each year in the United States. I get a pair of shoes or boots repaired about every four years. Assuming I am typical, the 2.3×10^8 people in the United States (1983) have about $2.3 \times 10^8/4$ or 5.75×10^7 repair jobs carried out each year. Since one cobbler can repair 4000 shoes in a year, we need $5.75 \times 10^7/4000$ or 14,375 cobblers to do all the repair work in the United States.

You should be careful not to write your answer as 14,375, how-ever. That number has five significant figures: the 1, the 4, the 3, the 7, and the 5, and a pretense to such accuracy is unjustified. An order-of-magnitude answer was wanted, so 10^4 will suffice. A more precise answer—one valid to five significant figures—would require input data precise to five significant figures, and we used no such data. Nonsignificant figures have a habit of accumulating in the course of a calculation, like mud on a boot, and you must wipe them off at the end. It is still good policy to keep one or two nonsignificant figures during a calculation, however, so that the rounding off at the end will yield a better estimate.

Try your hand at the exercises below. Provide order-of-magnitude answers.

EXERCISE 1: Suppose you have never watched a cobbler work, so you have no idea how long each job takes; but you have paid the cobbler's bill. How will you estimate the number of cobblers in the United States now?

EXERCISE 2: How many dentists are there in New York, a city of roughly 10^7 people? How many fresh tarts (along with cobblers and other fruit pastries) are there in the "Big Apple"?

EXERCISE 3: How many pairs of shoes can be made from a cow? (Hint: consider a spherical cow—and a spherical shoe, to boot.)

EXERCISE 4: About what fraction of a cubic centimeter of rub-ber is worn off an automobile tire with each revolution of the wheel?

2. Measuring Molecules

Benjamin Franklin dropped oil on a lake's surface and noticed that a given amount of oil could not be induced to spread out beyond a certain area.[3] If the number of drops of oil was doubled, then so was the maximum area to which it would spread. His measurements revealed that 0.1 cm^3 of oil spread to a maximum area of 40 m^2. How thick is such an oil layer?

· · · · · · ·

Let's denote the thickness of the layer by the symbol d. If d is expressed in units of meters, then the volume of that layer is $40d$ m^3. Since oil does not change volume much under changes in pressure or temperature, it is reasonable to assume that the volume of the oil sample does not change significantly simply by being spread out on a surface. Therefore, we can equate the initial volume, 0.1 cm^3, to the final volume, $40d$ m^3, and thus determine d. First, though, we must express both volumes in the same units. If we select cubic meters as our unit of volume, then we have to express 0.1 cm^3 in m^3. Since 1 m = 100 cm, it follows by cubing both sides that 1 m^3 = $(100)^3$ cm^3 or 1 m^3 = 10^6 cm^3. Hence, 1 cm^3 = 10^{-6} m^3 and 0.1 cm^3 = 10^{-7} m^3. Now that the units are consistent, we can equate $40d$ m^3 with 10^{-7} m^3 to get $d = 10^{-7}/40 = 25 \times 10^{-10}$, in units of meters.

A length of 10^{-10} m is called an angstrom and is denoted by the symbol Å. Thus, d equals 25 Å. The angstrom is a convenient unit because the lighter atoms such as hydrogen, carbon, and oxygen are on the order of magnitude of 1 Å in diameter. The distance between atoms in the molecules of common liquids and solids is also on the order of magnitude of 1 Å. The oil layer, then, is on the order of magnitude of ten atoms thick. For the kind of oil Franklin used, this is equivalent to being approximately one molecule thick. That is why such thin oil layers are called "monomolecular layers," and it is also why the oil layer would not spread thinner.

The pragmatic Franklin was interested in these experiments because he wished to explore the use of oil to calm rough waters and thereby prevent wave damage to ships. In Franklin's time, no one knew about molecules, but his creative experimental approach enabled him to make, in effect, the first estimate of a molecule's size!

EXERCISE 1: Franklin actually showed that 1 teaspoon of oil would spread to cover about 0.5 acre. Using the information in the

---

3. See Goodman (1956) for more on Franklin's many scientific achievements.

Appendix (I.3) that $10^4 \ m^2 = 2.47$ acres, determine how many cubic centimeters there are in a teaspoon.

EXERCISE 2: Estimate the average spacing between H_2O molecules in liquid water by making use of two pieces of information: (*a*) liquid water has a density of 1 g/cm^3, and (*b*) every 18 g of water contain Avogadro's number (6.02×10^{23}) of H_2O molecules.

3. The Size of an Ancient Asteroid

It has been proposed that dinosaurs and many other organisms became extinct 65 million years ago because Earth was struck by a large asteroid (Alvarez et al. 1980). The idea is that dust from the impact was lofted into the upper atmosphere all around the globe, where it lingered for at least several months and blocked the sunlight reaching Earth's surface. On the dark and cold Earth that temporarily resulted (Pollack et al. 1983), many forms of life then became extinct. Available evidence (Alvarez et al. 1980) suggests that about 20% of the asteroid's mass ended up as dust spread uniformly over Earth after eventually settling out of the upper atmosphere. This dust amounted to about 0.02 g/cm^2 of Earth's surface. The asteroid very likely had a density of about 2 g/cm^3. How large was the asteroid?

. . . .

To solve this problem we proceed in two steps. First we estimate the mass of the dust, and then we determine how big the asteroid must have been to contain that mass. The dust surrounds the Earth. According to the Appendix, Earth has an area of 5.1×10^{14} m^2, or 5.1×10^{18} cm^2. Since every square centimeter contained 0.02 g of dust from the asteroid, the dust layer contained a mass of (0.02 g/cm^2) $(5.1 \times 10^{18}$ cm$^2) = 1.02 \times 10^{17}$ g.

This much dust is 20% of the mass, M, of the asteroid, so the asteroid had a mass of

$$M = \frac{1.02 \times 10^{17}}{0.20} \text{ g} = 5.1 \times 10^{17} \text{ g.} \tag{1}$$

Now, consider a spherical asteroid with a radius, R. Its volume, V, is given by

$$V = \frac{4}{3} \pi R^3. \tag{2}$$

The mass of material in the asteroid is equal to the density, ρ, times the volume, or

$$M = \rho V = \rho \frac{4}{3} \pi R^3. \tag{3}$$

Using the fact that the density is 2 g/cm^3 and equating the expressions for M in Eqs. 1 and 3, we get

$$5.1 \times 10^{17} \text{ g} = 2 \frac{\text{g}}{\text{cm}^3} \times \frac{4}{3} \pi R^3. \tag{4}$$

Hence,

$$R^3 = \frac{5.1 \times 10^{17}}{(2)\left(\frac{4}{3}\pi\right)} \text{ cm}^3 = 0.61 \times 10^{17} \text{ cm}^3. \tag{5}$$

R, in units of centimeters, is then the cube root of 0.61×10^{17}. An approximate value for the cube root can be obtained by first noting that 0.61×10^{17} equals 61×10^{15}. The cube root[4] of 10^{15} is 10^5 and the cube root of 61 is about 4 (4^3 equals 64). So, roughly, R equals 4×10^5 cm or 4 km. Given the rounded-off estimates that went into the problem, it suffices to say that the diameter of the asteroid was about 10 km.

Additional dust would be produced from the earthly material that was blown out by the impact. It has been estimated that about 60 times as much material would be blasted out of the crater zone as was contained in the asteroid itself and that about 20% of this material from the crater would also be lofted to the upper atmosphere.

EXERCISE 1: If the asteroid created a crater 200 km in diameter, what would be its average depth? Assume that the factor of 60 in the last paragraph above is correct and that the density of the earthly material is also 2 g/cm^3.

EXERCISE 2: (*a*) If asteroids as big as 10 km in diameter or larger strike Earth at random times, but on the average once every 100 million years, what are the odds that one will hit during the next 50 years? (*b*) Suppose, now, that these asteroids strike random locations on Earth each time. If the next impact occurs tomorrow and results in a crater 200 km in diameter, what are the odds that you will be somewhere in the crater zone when the asteroid hits? You don't need fancy probability theory to answer these; just use the same common sense that tells you the odds are 1/52 that if you draw one card at random from a deck, it will be the ace of spades.

EXERCISE 3: Use a table of logarithms or a calculator to calculate $(61)^{1/3}$ to three significant figures.

---

4. In general, $10^a 10^b = 10^{a+b}$ and $(10^a)^b = 10^{ab}$ for any values of a and b. Therefore, since the cube root is designated by an exponent of 1/3, using the second of these relations, $(10^{15})^{1/3}$ equals $10^{15/3}$ or 10^5.

EXERCISE 4: When you look at a sphere of radius R, the area blocked by the sphere, called the cross-sectional area, is πR^2. The amount of sunlight blocked by the dust will depend on the summed cross-sectional area of all the dust particles rather than on their total mass. After all, if all the dust were lumped in one large particle up in the atmosphere it would block only a tiny portion of the total sunlight reaching Earth's surface. To explore this, assume each dust particle is spherical, with a density of 2 g/cm^3. (*a*) What would be the sum of the cross-sectional areas of all the dust particles lofted to the upper atmosphere from asteroid and crater if each dust particle had a radius of 5×10^{-7} m? (*b*) And if the radius were 10^{-5} m? Compare those areas with the surface area of Earth. Even though there will be overlapping particles, would you expect the dust to block a large fraction of incoming sunlight?

4. Exhausting Fossil Fuel Resources (I)

At the 1980 global consumption rate of petroleum, how long will it take to use up the estimated worldwide resource of this fuel?

· · · · · · ·

If a resource is used at a constant rate, year after year, then to determine how many years the resource will last, divide the amount of resource left by the annual rate of consumption. Using data in the Appendix (VII.2), the liftetime, T, of Earth's petroleum resources is

$$T = \frac{\text{quantity of resource}}{\text{rate of consumption in 1980}}$$

$$= \frac{1.0 \times 10^{22} \text{ J}}{1.35 \times 10^{20} \text{ J/yr}} \tag{1}$$

$$= 74 \text{ yr.}$$

Problem II.22 also deals with the exhaustion of energy resources, but under the conditions of exponentially growing consumption rather than constant consumption. Readers interested in the magnitude of Earth's energy resources will find the classic article by Hubbert (1969) most rewarding.

EXERCISE 1: What are the lifetimes of Earth's coal and natural gas resources at present (1980) consumption rates?

EXERCISE 2: Rivers presently carry about 10^{10} m^3 of soil and rock to the sea each year throughout the world. Roughly how long will it take at that rate for the continents to shrink by an average of 1 m in elevation if all other processes (such as continental uplift) are ignored?

5. Getting Denser

If the global human population continues to grow at the rate it averaged between 1950 and 1980, how long will it take for the average human population density on Earth's land to equal the present population density in typical urban areas of the world?

· · · · · · · ·

Although the problem does not state that the human population has been growing exponentially, this is a reasonable starting assumption. Indeed, given any set of data characterizing the human population over time—whether it concerns population, food consumption, energy consumption, industrial output, etc.—the exponential growth assumption will produce a good first approximation of how those data behave.[5] Hence, let's assume that between 1950 and 1980 the population, $N(t)$, behaved as

$$N(t) = N(0) e^{rt}, \tag{1}$$

where t is time, $t = 0$ is 1950, $N(0)$ is the population in 1950, and r is a parameter called the rate constant. With the aid of tabulated data in the Appendix (XVI) on the human population during this period, the validity of the exponential-growth assumption can be checked.

To begin, take the natural logarithm (logarithm to the base e) of Eq. 1:

$$\log_e[N(t)] = \log_e[N(0)] + rt. \tag{2}$$

Eq. 2 tells us that if $\log_e[N(t)]$ is plotted as a function of t, then the relation between $\log_e N$ and t is that of a straight line with slope r. If the data do, indeed, fall on such a straight line (at least approximately) then the growth is (approximately) exponential and the rate constant, r, can be determined from the graph. The graph of $\log_e[N(t)]$ versus t is shown in Figure I-1. A slope of about .019/yr is obtained. Thus the rate constant for human population growth is about 1.9% per year.[6]

---

5. Regrettably, our collective ability to cope with and control the consequences of our population growth seems not to increase with these other parameters.
6. If $r << 1$, e^r is about equal to $1 + r$. In that case population at the end of a year is about $(1 + r)$ times the population at the beginning of the year. In other words, the fractional increase in the population in one year is approximately equal to the rate constant, r. On the other hand, if r is not very small, then the fractional increase in population in one year can be significantly greater than r. When $r = 0.0190$, $e^r = 1.0192$, so the annual fractional increase in the population is nearly equal to r.

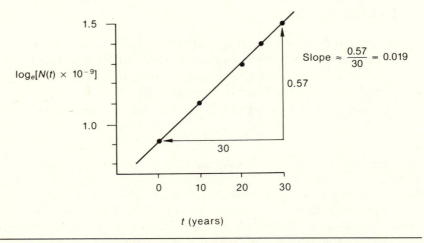

Figure I-1 The logarithm to the base e of 10^{-9} times the world's human population plotted as a function of time. The best straight-line fit to the data has a slope of 0.019; $t = 0$ corresponds to the year 1950 and $t = 30$ corresponds to 1980.

With r determined, all that is needed now is an estimate of the population density in a typical urban area. This will have to be guessed. There is no "correct" answer, since the concept of urban area is a fuzzy one. Very large cities contain on the order of 10^7 people and occupy perhaps 10^3 km^2, so let's take the urban density to be 10^4 people/km^2. If the total land area of Earth (1.5×10^8 km^2) is potentially accessible, then the estimated 1980 population of 4.5×10^9 people dwelled at an average density of about 30 people/km^2.

Because land area is fixed, density grows at the same rate as population. Therefore, we need only calculate how long it will take for 30 (present density) to increase exponentially to 10^4 (urban density) at a rate constant of 0.019/yr. Letting T denote the time period in question, we must solve the equation:

$$10^4 = 30 \, e^{rT}, \tag{3}$$

where $r = 0.019$/yr. Rearranging this equation,

$$e^{rT} = \frac{10^4}{30}. \tag{4}$$

Taking the natural logarithm of each side,

$$rT = \log_e \frac{10^4}{30}. \tag{5}$$

Hence, using a table of logarithms or a calculator,

$$rT = 5.8 \tag{6}$$

or

$$T = \frac{5.8}{r} = \frac{5.8}{0.019/\text{yr}} = 305 \text{ yr.} \tag{7}$$

Therefore, in the year 2285, the urban density would be attained worldwide.

EXERCISE 1:   What will be the world's human population in the year 2285 assuming a growth rate of 1.9% per year?

EXERCISE 2:   Often it is useful to know what the logarithm of 2 or e or 10 is to each of the bases 2, e, and 10. For example, knowing that a population grows exponentially as $N(0)\,e^{rt}$, the doubling time T_2, is given by $e^{rT_2} = 2$, or $T_2 = (\log_e 2)/r$. Sometimes one wants to know how many times a small number must double to reach a large number. Then it is useful to be able to express a number like 10^8 as a power of 2. If $2^x = 10^8$, $x = \log_2 10^8 = 8 \log_2 10$. Hence, the log of 10 to the base of 2 is needed. Fill in the rest of the table below and keep it handy for later use:

	2	e	10
$\log_2$	1		
$\log_e$		1	
$\log_{10}$			1

EXERCISE 3:   How many times will the population have doubled between 1980 and 2285 under the exponential growth assumption?

EXERCISE 4:   If a population, $N(t)$, is growing at a rate of 50% per year (i.e., the population at the end of each year is 50% greater than it is at the beginning of the year) and if the population, $N(t)$, is written as $N(t) = N(0)\,e^{rT}$, what is the value of r in units of yr^{-1}?

EXERCISE 5:   Plot the world population data from 1950 to 1980 on semilog graph paper and determine from the slope of the graph the value of the parameter, s, where $N(t) = N(0)\,10^{st}$. Explain how the value of s could be determined from the knowledge that $r = 0.019$ with the aid of one of the entries in the table above.

* *EXERCISE 6:*   Consider the world's human population growth between 1900 and 1980. First, plot the population data from 1900 to 1980 just as we did in Figure I-1. What value of r do you get? Is the

exponential assumption a good one over this 80-year period? Examination of your figure should convince you that the exponential function is not an accurate description of the global population data. Analysts of data often like to find a function that does describe the data accurately because the nature of that function may provide insight into the mechanisms that cause the data to change in time. Moreover, with such a function, it is tempting to project forward in time and thus gain some insight into what the future will look like. As a possible improvement on the exponential function, consider a general time-dependence of the population data of the form $N(t) = (at + b)^n$, where a, b, and n are constants. If a function of this form is going to fit the data well, then a plot of $[N(t)]^{1/n}$ versus t should look like a straight line with a slope of a and an intercept of b. For the cases $n = -1, 1, 2,$ and 3, plot on graph paper $[N(t)]^{1/n}$ versus t and determine by inspection which gives the best fit. What is shocking about your answer? Do you see why the temptation to project a good fit forward in time should sometimes be resisted?

6. The Greens We Eat

What fraction of the total annual plant growth on
Earth was eaten by humans in 1983?

· · · · · · ·

Just for fun, take a guess and write your guess down. Guessing
things like this and then later comparing your guess with the result
of a quantitative derivation helps sharpen your intuition. Now let's
proceed to the derivation.

The fraction we are looking for has annual human food consumption as its numerator and annual global green plant production as its
denominator. You must make several choices before you can calculate
this fraction.

First, choose your units. You can determine the numerator and
denominator in units of heat energy (e.g., calories) or in grams of
carbon, dry-weight biomass, or wet-weight biomass. Of course, if you
convert properly, you should get about the same answer using any
of these units.

Second, decide what is meant by annual plant growth. Will you
take the green-plant production rate to be the gross primary productivity (total photosynthetic activity) or the net primary productivity
(gross productivity minus losses due to plant respiration)? The answer will depend on which choice you make.

Third, be specific about the interpretation of human food consumption. Specifically, you can count meat consumption on the same
caloric or weight basis as plant matter, or you can estimate how much
plant matter it took to produce a unit of meat matter.

We will solve the problem using energy units and *net* primary productivity (npp). Net productivity is clearly the more sensible denominator because photosynthetic product respired by plants is of no benefit to plant eaters. We will not worry for now about whether human
energy needs are met by meat or green plants. (That issue is taken
up in Exercise 2.) Here we will simply compare the total food energy
consumed per year to the npp in units of energy per year.

The annual energy content of the typical human diet is given in
the Appendix (XV), although you can probably derive it from something you may know, namely, that the average person consumes
about 2500 Calories per day.[7] Calories with a capital "C" are weight-watchers' calories. One Calorie equals 1000 physicists' calories (lower

---

7. This figure represents a rough average of the food energy consumption in the underdeveloped nations, where many people have access to fewer Calories each day, and
in the overdeveloped nations, where many people eat more than 2500 Calories per day.

case "c").[8] Therefore, the average person consumes about 2.5×10^6 cal/day or 9×10^8 cal/yr. Multiplying by the present (1983) estimated world population of 4.7×10^9 people, we arrive at an annual rate of human food consumption of about 4.2×10^{18} cal/yr. In this book, and in most of the scientific literature today, the preferred energy unit is the joule (J). Using the conversion tables in the Appendix (I.8), we can express our numerator as 1.8×10^{19} J/yr.

The denominator, net primary productivity (in units of joules per year), can be obtained from information given in the Appendix (XII.1). Net primary productivity (in units of grams of carbon per year) is given as 7.5×10^{16} g(C)/yr.[9] This is the net amount of carbon converted from CO_2 to carbon-containing organic molecules each year. It includes photosynthate grazed by herbivores or detritivores subsequent to formation but does not include photosynthate that the plant itself burns for metabolic purposes. The Appendix also notes (VII.4) the energy content of dry biomass—about 1.6×10^4 J/g(biomass).

With npp in units of g(C)/yr and energy content of dry biomass in units of J/g(biomass), we can express npp in the same units as the numerator, J/yr, provided we can relate grams of dry biomass to grams of carbon. Thus, we need to know the fraction, by weight, of carbon contained in biomass, and then we can use the unit conversion formula:

$$\text{npp (J/yr)} = \frac{\text{npp [g(C)/yr]} \times \text{energy content [J/g(biomass)]}}{\text{carbon content [g(C)/g(biomass)]}}. \quad (1)$$

The carbon content of dry biomass can be estimated by looking at a typical plant compound and using its fractional carbon content as an approximation. Glucose is a common product of photosynthesis. Its chemical formula is $C_6H_{12}O_6$, meaning simply that a molecule of glucose contains 6 atoms of carbon, 12 of hydrogen, and 6 of oxygen. Exercise 1 provides a more accurate chemical representation of the atomic constituents of dry biomass, but here we can use the formula for glucose as an approximation. The molecular mass of glucose is $6 \, m(C) + 12 \, m(H) + 6 \, m(O)$, where $m(X)$ is the atomic mass of element X. Thus the molecular mass of glucose is $6(12) + 12(1) + 6(16) = 180$ (see Figure I-2). The molecular mass of the 6 carbon atoms in glucose is $6 \, m(C) = 72$, from which we deduce that

$$\text{carbon content} = \frac{g(C)}{g(\text{biomass})} = \frac{72}{180} = 0.4 \quad (2)$$

---

8. The physicists' calorie denotes the amount of heat needed to raise a gram of water from 14.5°C to 15.5°C.

9. The Appendix (I.1–14) also lists the unit abbreviations (e.g., g for grams) used throughout the book.

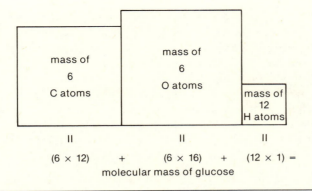

Figure I-2   The molecular mass of glucose ($C_6H_{12}O_6$) is calculated by adding together the mass of 6 carbon atoms (6 × 12), 6 oxygen atoms (6 × 16), and 12 hydrogen atoms (12 × 1).

Substituting Eq. 2 into Eq. 1, and using the above data from the Appendix,

$$\text{npp(J/yr)} = \frac{7.5 \times 10^{16} \text{ g(C)/yr} \times 1.6 \times 10^4 \text{ J/g(biomass)}}{0.4 \text{ g(C)/g(biomass)}} \tag{3}$$

$$= 3.0 \times 10^{21} \text{ J/yr.}$$

Putting all of the above together, the fraction, f, of npp consumed by humans is given by

$$f = \frac{\text{rate of human food consumption (J/yr)}}{\text{npp (J/yr)}}$$

$$= \frac{1.8 \times 10^{19} \text{ J/yr}}{3.0 \times 10^{21} \text{ J/yr}} \tag{4}$$

$$= 0.006.$$

In words, the rate at which energy is consumed by humans as food is about 0.6% or $1/160$ of the net rate at which energy is incorporated as plant matter in photosynthesis. How close was your guess?

EXERCISE 1:   A formula representing the approximate chemical composition of typical dry freshly photosynthesized biomass is H_{2960} O_{1480} C_{1480} N_{160} P_{18} S_{10}, where each subscript denotes the relative number of atoms of that elemental type. If this more precise representation is used instead of $C_6H_{12}O_6$, recalculate the fraction, f.

EXERCISE 2:   The production of animal-derived foods, such as beef, eggs, fish, and milk, requires the production of plants as fodder.

To produce 1 J of energy in the form of beef requires about 8 J of energy in the form of grains, while for poultry about 3 J of energy from grains are required. These represent extremes. The production of 1 J of other animal-derived foods requires very roughly 5 J of plant matter. Estimate how much meat you eat per year and use this to work out the following: If all of Earth's people ate a diet like yours, approximately what would the fraction f be? (Hint: if you start with an estimate of the mass of meat you eat, you will have to assume something about its water content. You may assume fresh meat has about the same water content as fresh vegetation—roughly 70%.)

EXERCISE 3: About what fraction of Earth's current npp would we need to consume if we derived all the energy we now (1980) get from fossil fuel from biomass instead? What does your answer tell you about the wisdom of replacing all fossil fuels with biomass? What ecological problems would you anticipate this might cause?

EXERCISE 4: If the human population continues to grow at about 2%/yr, in what year will humans be eating at Earth's current rate of npp?

7. Sulfur in Coal

How many tonnes and how many moles of sulfur were contained in the coal consumed worldwide in 1980?

$$\cdots \cdots \cdots$$

Lest you think this is just a bothersome question to keep you busy, read on. Sulfur in coal is at the root of a major environmental issue of our time—acid precipitation. In later problems, we will follow the track of sulfur in fossil fuel as it goes up into the atmosphere as a gas and comes back down as sulfuric acid.

In Section VII of the Appendix (VII) you will find the energy content of coal, the amount of energy derived from coal in 1980, and the sulfur content (by mass fraction) of typical coal burned throughout the world. Before worrying about specific numbers, however, it will help to write a general formula for the tonnes of sulfur in the coal.

Because the sulfur content of coal is expressed as a *mass* fraction (the mass of the sulfur in the coal divided by the mass of the coal) in the Appendix (VII.3), we must first determine the *mass* of coal burned per year. This can be done as follows:

$$\text{metric tons (tonnes) of coal consumed in 1980} = \frac{\text{energy derived from coal in 1980 (J)}}{\text{energy content per unit mass of coal (J/tonne)}} \quad (1)$$

Each quantity is followed by its representative units placed in parentheses. Note that the joules unit (J) cancels out in the numerator and denominator, leaving the metric tons unit (tonnes) on both sides of the equation.

The number of metric tons of sulfur in the coal can then be computed by substituting the above into the following formula:

$$\text{tonnes of sulfur from coal combustion in 1980} = \text{tonnes of coal consumed in 1980} \times \text{sulfur fraction of coal.} \quad (2)$$

Substituting numbers from the Appendix (VII.2, 3),[10]

---

10. Sometimes, when several chemical forms of a substance exist, there can be ambiguity about which form is intended. Thus, for example, emissions of sulfur to the environment can be expressed as tonnes of sulfur, but quite frequently are expressed as tonnes of sulfur dioxide, SO_2. A tonne of SO_2 contains only half a tonne of sulfur (because the molecular mass of S and the molecular mass of O_2 are equal), and therefore the distinction is important. Symbols such as tonne(S) or tonne(SO_2), rather than plain tonne, are commonly used to avoid ambiguity and will be used throughout this book.

tonnes of sulfur from coal combustion in 1980

$$= \frac{90 \times 10^{18} \text{ J}}{29.3 \times 10^9 \text{ J/tonne(coal)}} \times 0.025 \text{ tonnes(S)/tonne(coal)} \tag{3}$$

$$= 7.7 \times 10^7 \text{ tonnes(S)}.$$

A mole of any material has a mass of M grams, where M is the molecular mass of the substance (number of protons and neutrons in a molecule). The following units conversion is used to convert tonnes to moles for any material, **A**:

$$\text{moles of } \mathbf{A} = \frac{\text{tonnes}(\mathbf{A}) \times [10^6 \text{ g}(\mathbf{A})/\text{tonne}(\mathbf{A})]}{M[\text{g}(\mathbf{A})/\text{mole}(\mathbf{A})]} \tag{4}$$

Using the fact that the atomic (and molecular) mass[11] of S is 32, and substituting the actual numbers into Eq. 4, we get:

moles(S) from coal combustion in 1980

$$= \frac{[7.7 \times 10^7 \text{ tonnes(S)}] \times [10^6 \text{ g(S)/tonne(S)}]}{32 \text{ g(S)/mole(S)}} \tag{5}$$

$$= 2.4 \times 10^{12} \text{ moles(S)}.$$

EXERCISE 1:   Using additional data from the Appendix (VII.3), calculate the number of tonnes of cadmium, lead, zinc, selenium, mercury, and arsenic contained in the coal combusted worldwide 1980. Compare the rates at which these substances are emitted into the atmosphere worldwide from coal combustion with the natural background rates of mobilization to the atmosphere given in the Appendix (IX), under the assumption that all the trace substances found in fossil fuel are emitted into the atmosphere when the fuel is burned. In practice, some portion of the trace substances in fuels is left behind as solid waste in the form of ash when the fuel is burned. The pathways that the substances in fuel follow (through air, water, soil, and living organisms) and the chemical transformations they undergo in the environment (including reactions that can render them more or less toxic than they were in their initial emitted form) depend on many factors. Specific examples will be described later in this book.

EXERCISE 2:   For further practice at converting units, use the data in the Appendix on the energy content of specified quantities of fossil fuels to calculate the present world's resources of petroleum, coal, and natural gas in units of barrels, metric tons, and cubic meters, respectively.

---

11.   For a substance like molecular oxygen (O_2) consisting of two identical atoms, the molecular mass is twice the atomic mass; but for S or any other molecule consisting of a single atom, the molecular mass equals the atomic mass.

Chapter II
Tools of the Trade

. . . having a good question, a fundamental question, and
having some tools of inquiry that allow you to take the first
step toward an answer—those are the conditions that make
for exciting science.

—Herbert A. Simon

Here you will be handed some of the tools that form the core of environmental science. They include residence-time methods and box
models, practical methods in thermodynamics and chemical equilibrium kinetics, and a few relatively simple differential equations. (If
you are typical, you were probably paralyzed with fear when you
read the previous sentence. Relax—we will take it slowly.)

A.
Steady-State Box Models and Residence Times

When the flow of a substance into a lake, the atmosphere, an animal, or any other "box" is equal to the outflow of that substance, then the amount, or "stock," of that substance in the box will be constant. This is called a "steady state" or "equilibrium." The ratio of the *stock* in the box to the *flow rate* (in or out) is called the *residence time*. Thus, if F_{in} is the rate of inflow to the box and F_{out} is the rate of outflow, the steady-state condition is $F_{in} = F_{out}$. Letting M be the stock and T be the residence time, $M/F_{in} = M/F_{out} = T$. The first problem provides a very simple illustration of the concept of residence time, while the subsequent problems provide practice in applying it. Problems 2 through 5 in this chapter are "one-box" problems, with only a single inflow and a single outflow to worry about in each. Problem 6 is also a "one-box" problem with single inflow and outflow, but the discussion following the solution describes how to deal with multiple flows. In Problem II.7, two boxes are interconnected by flows. Problems II.8 and II.9 also involve connected boxes and treat the case in which a flow or stock is perturbed. The last two problems illustrate the widespread applicability and power of box models. Problem II.10 shows how flow-stock considerations allow estimation of the deposition velocity of airborne particulates, and Problem II.11 shows how a steady-state indoor air concentration of a radioactive isotope can be derived. Later problems in Chapter III illustrate how thinking in terms of flows, stocks, and residence times can help simplify situations more complicated than those treated here.

1. School as a Steady-State System

A college has a constant undergraduate enrollment of 14,000 students. No students flunk out or transfer in from other colleges and so the residence time of each student is four years. How many students graduate each year?

. . . .

The residence time of a student is four years, the stock of students is 14,000, and the outflow rate is the graduation rate. It follows from the preceding discussion of steady-state systems that the flow rate in or out of the college is the stock divided by the residence time, or

$$F_{in} = F_{out} = M/T. \tag{1}$$

For our specific problem, this formula becomes:

$$\text{graduation rate} = \frac{\text{total stock of students}}{\text{residence time of students}} = \frac{14,000}{4 \text{ yr}}. \tag{2}$$

$$= 3,500/\text{yr.}$$

In this problem, all students spend exactly the same amount of time in college—four years. In many other steady-state situations, each unit of the substance (for example, each molecule of pollutant) comprising the stock spends the residence time in the box only in a statistical sense. The average time spent by all the units is the residence time, but the individual times spent by the units may differ widely. Provided the stock is in a steady state, stock = (inflow or outflow rate) × (residence time). The flow can be of various types. The movement of students through college illustrates a type of flow in which the components of the stock pass through in an orderly manner so that each component has the same residence time. The subsequent box-model problems in this chapter illustrate the far more typical case of *mixed flow*, in which the inflowing units of stock mix thoroughly in a medium and have differing individual residence times.

EXERCISE 1: A population of cows on a farm is in steady state. The birth rate is 7 calves per year and the average residence time for a cow on the farm is 6 years. How large is the herd?

EXERCISE 2: Suppose there are 100 students enrolled in a graduate program year after year, and that each year 20 get degrees and leave, 5 flunk out, and 25 new students enter the program. (*a*) What is the average residence time of a student in the program? (*b*) If all the students that flunk out do so at the end of their first year, what is the average length of time to get a degree (for those students who do get a degree)? (Hint for *b*: First determine the number of enrolled students, at any specific time, who will eventually get degrees.)

2. The Water Above

What is the residence time of H_2O in Earth's atmosphere?

· · · · · · ·

The sky is often fairly clear after a big rain storm, so as a rough estimate, the average interval between major precipitation events should be the atmospheric residence time of water. Half of that interval might be a better guess, however, because evaporation and transpiration[1] rates are probably fairly steady during the between-storm interval (see Figure II-1). (Can you see why it makes no difference to this argument that precipitation does not fall synchronously all over the world?) This "hand-waving" argument suggests that the residence time should be very roughly a week, an answer that should underestimate the true solution because the sky is not actually free of all moisture after a storm.

An accurate answer can be obtained by appropriate use of data in the Appendix. Assuming the atmospheric H_2O is in steady state, the

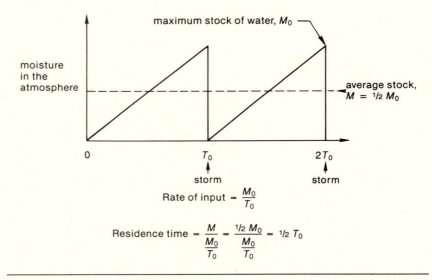

Figure II-1 An hypothesis about the buildup of atmospheric water between storms. This hand-waving argument suggests the residence time of atmospheric water is half the interval between storms.

---

1. The glossary provides definitions of words like "transpiration" that may be unfamiliar to you. If you come across a word you don't know, check the glossary first.

flow of H_2O into the atmosphere, F_w, equals the flow out. The outflow, of course, is the global precipitation rate. From the Appendix (VI.2,3) we learn that $F_w = 5.18 \times 10^{14}$ m^3/yr, and that the stock of H_2O in the atmosphere, M_w (which is mostly in vapor, not liquid, form), is 1.3×10^{13} m^3 (liquid equivalent). The residence time, T_w, is then

$$T_w = \frac{M_w}{F_w} = 0.025 \text{ yr} = 9.1 \text{ days}. \tag{1}$$

The residence time of individual water molecules in the atmosphere may range from fractions of an hour to millenia. T_w, calculated above, is the *average* residence time of all the water molecules in the atmosphere.

EXERCISE 1: What is the residence time of water in all of Earth's oceans?

EXERCISE 2: For further practice at converting units, calculate how much annual precipitation must occur on the average throughout the world, in units of inches of water, to account for the value of F_w.

3. Carbon in the Biosphere

What are the residence times of carbon in continental and marine vegetation?

. . . .

Take a guess. Trees live for 50 years or more and constitute a large share of continental vegetation. Phytoplankton, accounting for a large share of marine primary productivity, generally come and go in a series of annual blooms and crashes each lasting less than a year. We might therefore expect the residence time of carbon in plants to be on the order of 10–100 years on the land and 0.1–1 year in the oceans.

To be more accurate, we must know the stocks and flows of carbon. The stocks are living biomasses and the flows are net primary productivities. The Appendix provides both of these quantities, in conveniently comparable units—g(C) and g(C)/yr. From this we can obtain

$$
\begin{aligned}
T_{\text{terrestrial}} &= \frac{\text{stock of living continental biomass}}{\text{continental net primary productivity}} \\
&= \frac{5.6 \times 10^{17} \text{ g(C)}}{5 \times 10^{16} \text{ g(C)/yr}} \\
&= 11.2 \text{ yr}
\end{aligned}
\tag{1}
$$

and

$$
\begin{aligned}
T_{\text{oceanic}} &= \frac{\text{stock of living marine plants}}{\text{marine net primary productivity}} \\
&= \frac{2 \times 10^{15} \text{ g(C)}}{2.5 \times 10^{16} \text{ g(C)/yr}} \\
&= 0.08 \text{ yr} \approx 1 \text{ month.}
\end{aligned}
\tag{2}
$$

These relations are illustrated in Figure II-2. The considerable uncertainty in the stock and flow data used here (see Appendix) makes these residence times approximate.

Notice that the continental residence time of carbon is considerably shorter than the average lifetime of trees. Even though the woody parts of trees constitute the bulk of living continental biomass, only a part of each tree's annual production adds woody tissue; much of the net primary productivity produces leaves, which have less than a one-year residence time.

Why did we use net primary productivities in this calculation? Had we used data on gross marine and continental primary productivities

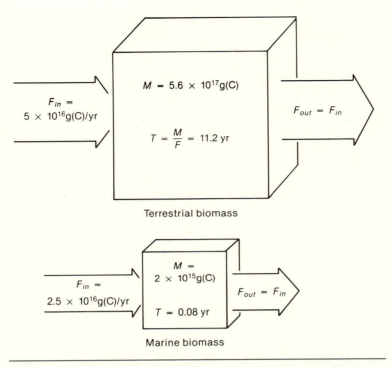

$M = 5.6 \times 10^{17} g(C)$

$F_{in} = 5 \times 10^{16} g(C)/yr$

$F_{out} = F_{in}$

$T = \dfrac{M}{F} = 11.2 \text{ yr}$

Terrestrial biomass

$M = 2 \times 10^{15} g(C)$

$F_{in} = 2.5 \times 10^{16} g(C)/yr$

$F_{out} = F_{in}$

$T = 0.08 \text{ yr}$

Marine biomass

Figure II-2 Stocks and flows of biomass in terrestrial and marine biomasses.

instead, we would have obtained much shorter residence times. Carbon flows relatively quickly through respiratory pathways in organisms. This respiratory flow is not included in net primary productivity. Using net primary productivity yields a residence time that bears a closer relation to the lifetimes of typical organisms (see Exercise 2).

EXERCISE 1: Suppose that the average residence time (ignoring respiratory pathways) of carbon in the phytoplankton in a lake is two weeks. Zooplankton in the lake, grazing upon the phytoplankton, consume 40% of the net primary productivity and have an incorporation efficiency of 25% (i. e., 25% of the phytoplankton biomass they eat is incorporated into zooplankton biomass). In other words, the net productivity of the zooplankton is 0.25 × 0.40 or 10% of the npp of the algae. If the average residence time (ignoring respiratory pathways) of carbon in zooplankton biomass is six months, estimate the ratio of the average biomass of the zooplankton population to that of the phytoplankton population in the lake. Figure II-3 illustrates the flows in and out of the two stocks of plankton.

* *EXERCISE 2:* In this exercise we explore the relation between residence time of biomass and lifetime of individuals in a steady-state

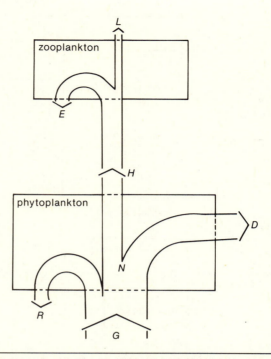

Figure II-3 The flows of carbon in and out of a phytoplankton and zooplankton population in a lake. G is the gross primary productivity of the phytoplankton and R is their respiration rate. $N = G - R$ is net primary productivity. H represents loss of carbon by zooplankton herbivory, and D comprises all other losses. E is excretory and metabolic loss of carbon from the zooplankton population, and $L = H - E$ describes the loss of carbon from the zooplankton population by predation or other forms of death. N and L are the flows used to determine the carbon residence time in the biomass of each population.

population. The residence time, T, of biomass in a population is the interval during which new net production of biomass equals the total average standing crop or stock, M. To relate T to lifetime of individuals, we must know something about the mass-versus-age dependence of the population. Consider two cases: (*a*) All individuals grow linearly in mass from birth until death. (*b*) All individuals achieve maximum mass early in life and then maintain a constant mass until death. The first case corresponds very crudely to the situation among plant populations, the second to that among animals. If it is assumed that all individuals in a population die at the same age, A_0, show that in case *a*, T equals $A_0/2$ and in case *b*, T is about equal to A_0.

* *EXERCISE 3:* Consider a forest in which the residence time for carbon (ignoring respiratory pathways) in the living trees is 15 years,

in leaves is 1 year, and in woody parts (including roots) is 100 years. (*a*) What is the ratio of leaf biomass to wood biomass, and (*b*) what fraction of total net productivity does wood productivity constitute?

4. Natural SO₂

Natural sources add sulfur dioxide (SO_2) to the atmosphere at a rate of about 10^8 tonnes(S)/yr. The background concentration of atmospheric SO_2, measured in remote areas where anthropogenic sources are not likely to have much influence, is about 0.2 parts per billion, by volume [ppb(v)]. What is the residence time of atmospheric SO_2 in the remote regions?

· · · · · · ·

The Appendix notes a total natural background sulfur emission rate of 1.5×10^{11} kg(S)/yr, or equivalently, 1.5×10^8 tonnes(S)/yr. Of this, roughly 10^8 tonnes(S)/yr comprise hydrogen sulfide (H_2S) and SO_2, while the rest is sulfate (SO_4^{-2}). The H_2S is rapidly converted to SO_2 in the atmosphere, whereas the SO_4^{-2} is not. Hence we estimate that 10^8 tonnes(S)/yr of SO_2 is emitted from natural sources.

Because the source term is in units of tonnes(S)/yr, it will be useful to convert the concentration given in the problem, 0.2 ppb(v), into units of tonnes(S). Since the (v) of ppb(v) stands for volume, and ppb stands for parts per billion (one part in 10^9), every unit volume of atmospheric air contains (on average) 0.2×10^{-9} unit volumes of SO_2. The use of volumetric comparisons of gases makes sense because of a special property of gases: Every mole (see Problem I.7) of gas occupies a volume of 22.4 liters at STP.[2] By comparing gases on a volume-per-volume basis, we are really comparing on a mole-per-mole basis.[3]

The number of moles of background SO_2 in our atmosphere can now be determined by multiplying the number of moles of air by the molar fraction of SO_2 calculated above (0.2×10^{-9}). So let's determine how many moles of air Earth's atmosphere contains. The mass of our planet's supply of air is 5.14×10^{21} g. To calculate the number of moles involved we must divide this by the molecular mass of air. Air, of course, is a mixture of gases. The molar fraction of the mixture that is nitrogen is 78.08%; the molar fraction that is oxygen is 20.95%. Because the molecular masses of N_2 and O_2 are 28 and 32, respectively, the average molecular mass of air is about $[(0.7808 \times 28) + (0.2095 \times 32)]/0.9903 = 28.85$. With trace gases included (see Appendix (V.2)), the average molecular mass works out to be 28.96. Dividing 5.14×10^{21} g of air by 28.96 g per mole of air yields a planetary total of about 1.8×10^{20} moles of air.

---

2. Standard temperature (0^0C) and pressure (1 atmosphere).
3. A mole of any substance has a mass equal to the molecular mass of the substance and contains 6.02×10^{23} molecules. The number 6.02×10^{23} is called Avogadro's number.

The product of moles of air times molar fraction of SO_2 gives $(0.2 \times 10^{-9}) \times (1.8 \times 10^{20})$ or 3.6×10^{10} moles of SO_2. Next, we must calculate the mass of this many moles of SO_2. The molecular mass of S is 32, so a mole of S contains 32 g of S. Therefore, the number of grams of S in atmospheric SO_2 is $32 \times 3.6 \times 10^{10} = 1.15 \times 10^{12}$ g(S) $= 1.15 \times 10^{6}$ tonnes(S).

We now have a natural background flow of SO_2 to the atmosphere, F, given by

$$F = 10^8 \text{ tonnes(S)/yr} \tag{1}$$

and a stock, M, given by

$$M = 1.15 \times 10^6 \text{ tonnes(S).} \tag{2}$$

The residence time, T, is equal to M/F, or

$$T = \frac{1.15 \times 10^6 \text{ tonnes(S)}}{10^8 \text{ tonnes(S)/yr}} = 0.0115 \text{ yr} \tag{3}$$
$$= 4.2 \text{ days.}$$

EXERCISE 1: Show that an SO_2 atmospheric concentration of 0.2 ppb(v) is equivalent to 0.4 ppb(w), where (w) stands for weight and ppb(w) refers to the number of tonnes (or grams) of SO_2 in 10^9 tonnes (or grams) of air.

EXERCISE 2: If gases emitted from Earth's surface to the atmosphere are, chemically, relatively nonreactive, and have low solubility in rain, they are likely to remain in the atmosphere for a long time. Under such circumstances, they have a good chance of passing through the tropopause to the stratosphere. A stable, nonreactive gas typically resides in the troposphere for about 10 years before exiting to the stratosphere. In the stratosphere, photochemical reactions may result in a sink, or exit process, for the gas. One such gas is nitrous oxide (N_2O), produced by bacteria in the natural denitrification process. Its chemical inertness gives N_2O safe passage through the troposphere, but in the stratosphere it is photochemically destroyed. The average concentration of N_2O in the troposphere is about 300 ppb(v). What is the global rate of production of N_2O in units of kg(N_2O)/yr?

EXERCISE 3: When gases are emitted to the atmosphere from Earth's surface, diffusion and atmospheric circulation cause the gases to mix. After about three weeks such gases are distributed vertically in the troposphere in such a way that their densities decrease with increasing altitude at the same rate as nitrogen's and oxygen's densities decrease. Qualitatively, how will the ratio of the SO_2 concentration to the N_2 or O_2 concentration depend on altitude within the troposphere?

5. Anthropogenic SO_2

With anthropogenic sources included, what is the globally averaged SO_2 concentration in the atmosphere? What is the SO_2 concentration in industrialized regions like the northeastern United States?

· · · · · · ·

Referring to the Appendix, anthropogenic sulfur emissions to the atmosphere in 1980 were about 8.5×10^7 tonnes(S)/yr.[4] Therefore, the globally averaged total SO_2 concentration will be about 85% higher than the result derived in Problem II.4, where only natural sources were included in the calculation. Anthropogenic emissions, however, will not be distributed as uniformly in the atmosphere as the background sulfur emissions because the sources are concentrated and the residence time is short compared to the time needed for gases to circle the globe.

Concentrations of atmospheric SO_2 over industrialized regions will vary tremendously, depending on weather conditions. Very importantly, it will also depend on the presence of other substances in the atmosphere that react with SO_2. These substances themselves are often the products of combustion and they can transform SO_2 to some other chemical state—for example, to sulfuric acid, which comprises acid precipitation. The occurrence of these reactions will alter the residence time of SO_2. The residence time approach will not yield a precise answer now, even if the above factors are specified, because the urban air box, or air shed, is not of precise size. Put differently, concentrations of pollutants trail away the further away one is from pollution sources, and the area they flow into and out of has no well-defined boundaries. However, a rough estimate of the concentrations in large industrialized regions like the northeastern United States can be obtained by noting that in four days of residence in the atmosphere, a typical SO_2 molecule will travel in the direction of a prevailing wind of 15 km/hr a distance of (15 km/hr) $\times$ (4 $\times$ 24 hr) = 1440 km. Thus, although wind velocity is variable, the northeastern section of the United States (from the latitude of Washington D.C. north, and from the midwest to the Atlantic coast) can be considered an approximate air shed for SO_2 emitted in that region (see Figure II-4).

The Northeast presently emits about 10^7 tonnes(S)/yr, or 12% of the global anthropogenic SO_2, yet it occupies only about 0.2% of Earth's

---

4. This is about 10% higher than the result for sulfur in coal found in Problem I.7; sulfur emissions result from both coal and petroleum combustion, but not all the sulfur in either fuel is emitted to the atmosphere.

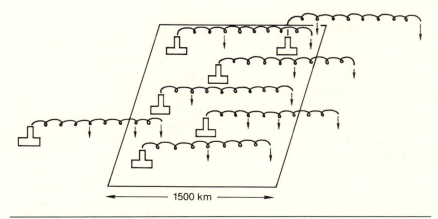

Figure II-4 Most, but not all, of the pollution that originates in a regional air shed will be deposited in it; most, but not all, of the pollution deposited in it also originated within its rough boundaries.

surface area. The SO_2 concentration in the region can be estimated as follows. The globally averaged SO_2 concentration due solely to anthropogenic SO_2 is roughly 85% of the natural background concentration, or 0.85×0.20 ppb(v) $= 0.17$ ppb(v). If this is multiplied by the fraction of anthropogenic SO_2 produced in the Northeast (0.12) and divided by the fraction of Earth's area occupied by this air shed (0.002), an approximate regional concentration of SO_2 is obtained of 10.2 ppb(v). To strive for precision, the natural background of 0.20 ppb(v) of SO_2 might be added to this value, although such a correction is lost in a sea of uncertainty.

EXERCISE 1:   We assumed, above, that the four-day residence time derived in Problem II.4 for remote sites was applicable to SO_2 in the atmosphere above the northeastern United States. If the residence time in dirty air were different from that in clean air, would you expect it to be longer or shorter? Why?

EXERCISE 2:   As a very rough estimate of the likely SO_2 concentration in the atmosphere above densely populated urban areas, one could divide the percentage of total SO_2 emitted in the city by the percentage of Earth's surface area the city occupies. If that ratio were multiplied by 0.17 ppb(v), as above, would the resulting estimate be a reasonable one? Would it underestimate or overestimate the true answer? After you have tackled Problems 1 and 9 in Chapter III, you might want to return to this urban SO_2 problem and develop a better model.

6. A Polluted Lake

A stable and highly soluble pollutant is dumped into a lake at the rate of 0.16 tonnes per day. The lake volume is 4×10^7 m^3 and the average water flow-through rate is 8×10^4 m^3/day. Ignore evaporation from the lake surface and assume the pollutant is uniformly mixed in the lake. What eventual steady-state concentration will the pollutant reach?

· · · · · · ·

The rate at which pollution is added to the lake is given, so to calculate the steady-state stock, the residence time is needed. Because the pollutant is uniformly mixed in the lake, the residence time of the pollutant will equal the residence time of the lake water, which can be derived from the lake data provided. Dividing the stock of water, M_w, by the rate of water flow-through, F_w, the residence time of water in the lake, T_w, is obtained:

$$T_w = \frac{M_w}{F_w} = \frac{4 \times 10^7 \text{ m}^3}{8 \times 10^4 \text{ m}^3/\text{day}} = 500 \text{ days}. \tag{1}$$

Because the residence time of the pollutant, T_p, is equal to T_w, it follows that the steady-state stock of pollutant, M_p, is the pollution input rate, F_p, times the residence time, or:

$$M_p = F_p T_p = 0.16 \text{ tonnes/day} \times 500 \text{ days} = 80 \text{ tonnes}. \tag{2}$$

If we multiply the volume of a cubic meter of water by the density of water, we discover that a cubic meter of water weighs exactly one metric ton. Thus, the steady-state concentration of the pollutant is 80 tonnes/(4×10^7) tonnes, or 2 parts per million (2.0×10^{-6}) by weight.

Aqueous concentrations are often specified in units of molarity, or moles per liter. Suppose the pollutant has a molecular weight of 40 (that is, there are a total of 40 protons and neutrons in the atoms of each molecule). Then the number of moles of pollutant is the weight in grams divided by 40, or $80 \times 10^6/40 = 2.0 \times 10^6$ moles. The number of liters of water in the lake is 4×10^{10}, so the molar concentration of the pollutant is 50×10^{-6} moles/liter. This is often written as 50 micromoles/liter, since a micromole is 10^{-6} moles. The unit of mole/liter is called a molar, and is sometimes designated by M. Thus the concentration can be expressed as 50×10^{-6} M or 50 μM.

Now let us look at our two important assumptions. First, suppose that evaporation cannot be ignored, so the total rate at which water exits the lake now has two components: evaporation (one third) and stream outflow (two thirds) (see Figure II-5). The total rate at which water exits the lake is unchanged; it equals the inflow rate of 8×10^4 m^3/day. Moreover, assume that the evaporating water is free of pollutant.[5] Qualitatively, we would expect the steady-state concentration of pollutant to be higher now, because one possible exit pathway (with evaporating water) is closed off. The residence time of the pollutant is now no longer equal to that of the water but rather is given by the residence time associated only with stream outflow of water. This is given by

$$T_{w,\,outflow} = \frac{4 \times 10^7 \text{ m}^3}{5.3 \times 10^4 \text{ m}^3/\text{day}} = 750 \text{ days.} \tag{3}$$

The rest of the calculation remains the same, and the steady-state concentration of pollutant will be 3/2 greater than before.

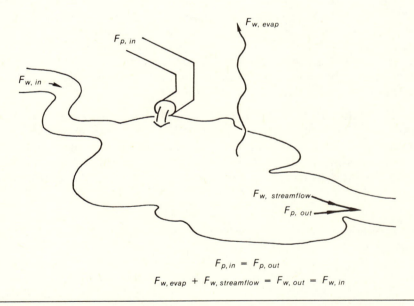

$$F_{p,\,in} = F_{p,\,out}$$
$$F_{w,\,evap} + F_{w,\,streamflow} = F_{w,\,out} = F_{w,\,in}$$

Figure II-5   The flows of pollution, F_p, and water, F_w, in and out of a polluted lake. It is assumed that the pollutant does not codistill with the evaporating lake water. The steady-state conditions for both pollutant and water are shown.

---

5.   A substance that leaves the water with the evaporating vapor is said to "codistill." Some pollutants, like DDT, do codistill, but most do not.

The other fundamental assumption made in the original problem was that the pollutant uniformly mixes in the lake water. The effect of relaxing this assumption will depend on precisely how the pollutant is unevenly distributed within the lake and how water flows through the lake. An important special case, for example, is that of a stratified lake, in which the upper, warmer layer (epilimnion) is relatively isolated from the lower, colder layer (hypolimnion). Stratification is fairly common in late spring, summer, and early fall in deep lakes in regions with distinct warm and cold seasons [see Wetzel (1975) for a complete discussion of this phenomenon]. A pollutant entering the epilimnion of a stratified lake will not mix readily with the hypolimnionic water. Therefore, the *effective* dilution volume is the volume of the epilimnion rather than that of the whole lake, provided the residence time of the pollutant is short compared to the time constant characterizing the exchange of water between the two stratified layers.

EXERCISE 1: Assume that there is no transfer of water between the two layers during six months of the year when the lake is stratified. During the remaining six months, the lake is thoroughly and rapidly mixed. Assume the epilimnion occupies one fifth of the volume of the lake and that all the inflowing water and the incoming pollutant enter the lake at the surface in such a manner that during stratification they mix only with the epilimnion. Sketch, qualitatively, a graph of the concentration of pollutant as a function of time throughout the year at two depths: one midway down the epilimnion and one midway down the hypolimnion.

EXERCISE 2: Two lakes are located on the same river, as shown in Figure II-6. A is upstream of B. Water flows into A at a rate S_A; it evaporates from A at a rate E_A and from B at a rate E_B. A tributary flows into the river between A and B at a rate S_B. Evaporation from the streams can be ignored. A soluble pollutant flows into lake

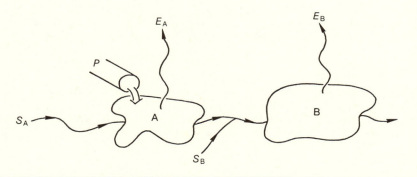

Figure II-6 Two lakes, A and B, on a stream. Lake A receives pollution directly, while B receives pollution from the outflow from A.

A at a rate P. There are no other sources of the pollutant, it is well mixed in both lakes, and it does not codistill. The lakes have water volume V_A and V_B, respectively, and are in hydrological steady states. (*a*) What is the rate of stream flow out of lake A? Into lake B? (*b*) What is the residence time of water in lake A? In lake B? (*c*) In the steady state, what concentration of pollutant will be found in each lake if P is given in units of grams per second, the S's and E's are in units of liters per second, and the V's are in units of liters?

EXERCISE 3: If a substance can leave a box by several distinct exit pathways, as water leaves a lake either by evaporation or in an outflowing stream, then individual residence times can be associated with each exit process. Let $F_{out} = F_{out,1} + F_{out,2} + \ldots + F_{out,n}$, and let the stock be M. Then the overall residence time, T, is equal to M/F_{out}. The individual residence times are given by $T_1 = M/F_{out,1}$, $T_2 = M/F_{out,2}, \ldots, T_n = M/F_{out,n}$. Prove that $T^{-1} = T_1^{-1} + T_2^{-1} + \ldots + T_n^{-1}$, and give an intuitive explanation of the meaning of these individual time constants. Is it correct to say that a molecule of the substance exiting by the ith pathway will, on the average, have spent a time period T_i in the box before exiting? In Exercise 3 of Problem II.3, the residence times for carbon in leaves ($T_L = 1$ yr), wood ($T_W = 100$ yr) and the trees as a whole ($T_T = 15$ yr) do not satisfy the relation $T_T^{-1} = T_L^{-1} + T_W^{-1}$. Why not?

7. The Flow of Atmospheric Pollutants between Hemispheres

Ethane (C_2H_6) is a constituent of natural gas. It is emitted to the atmosphere whenever natural gas escapes unburned at wells and other sources, a process that constitutes the only major source of ethane in the atmosphere. The average concentration of ethane in the troposphere of the northern hemisphere, C_N, is roughly 1.0 ppb(v), and the average concentration in the southern hemisphere, C_S, is roughly 0.5 ppb(v). Ethane can exit from the troposphere by any of three mechanisms: passage to the stratosphere; chemical reaction resulting in transformation to other chemical species; and deposition to Earth's surface (for example, by washout from the atmosphere in rain or snow). It can also leave one hemisphere's troposphere by flowing to the other's. Assuming that the total exit rate from each hemisphere's troposphere is proportional to the concentration in the respective troposphere, and knowing that 3% as much natural gas escapes to the atmosphere unburned as is burned, estimate the net rate of ethane flow across the equator.[6]

. . . .

The data in this problem suggest a model with two boxes—one each for the atmospheres of the northern and the southern hemispheres. In problems of this type, a decision has to be made about whether to use as variables in the model the *amounts* or the *concentrations* of the substance studied. The two measures are interconvertible (amount equals concentration times the constant volume or mass of the medium) and so the correctness of the final answer will not depend on this decision, which is solely a matter of convenience. Inflows or emissions of substances are usually stated in units of amount per unit time, though, so I prefer to use amounts, not concentrations, as variables.

Figure II-7 shows the two tropospheric boxes with inflows, outflows, and stocks labeled. X_N and X_S refer to the amounts (or stocks) of ethane in the two boxes. Inflows E_N and E_S comprise ethane emis-

---

6. The values for C_N and C_S come from Singh et al. (1979). This problem illustrates how just a few experimental numbers can be used to estimate a very important parameter in environmental science—the interhemispheric flow of trace substances in the troposphere.

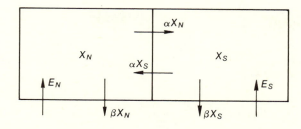

Figure II-7 The inflows and outflows of ethane in a two-box model of Earth's atmosphere. X_N and X_S are the amounts of ethane in the atmosphere above the northern and southern hemispheres, respectively. E_N and E_S are source terms, αX_N and αX_S describe flow across the equator, and βX_N and βX_S describe sinks within each box.

sions to the boxes from the ground. βX_N and βX_S are the total outflow rates from each box resulting from the three exit mechanisms mentioned in the problem statement. Finally, Figure II-7 separates the net interhemispheric flow into its two gross components—a flow from north to south labeled αX_N and a flow from south to north labeled αX_S. The net flow from north to south is the difference $\alpha(X_N - X_S)$.

The outflow rates, βX_N and βX_S, and the interhemispheric flow rates, αX_N and αX_S, are examples of "linear, donor-controlled flows" in which a rate of flow is simply proportional to the stock from which the flow originates. Intuition usually tells you when a flow is of this kind. In the case of deposition from the atmosphere to the ground, for example, each molecule of ethane falls independently of all the other molecules; thus the rate of deposition by fallout or washout when n molecules are present in the atmosphere ought to be half as great as that when $2n$ molecules are present. For ethane removal by chemical reaction, the situation is more complex. For example, if ethane reacts with a trace substance in the atmosphere, and if that substance is in short supply, then doubling the ethane concentration will not necessarily lead to a doubling of the reaction rate (see Exercise 6).

The convecting air that crosses the equator carries with it an amount of pollutant proportional to the amount of pollutant in the donor hemisphere. Thus, interhemispheric mixing is certainly a linear, donor-controlled process. Exit to the stratosphere should also be linear and donor-controlled. However, the inflow rates, E_N and E_S, are generally not proportional to the stocks. Polluters do not pollute in proportion to the amount of pollution present (at least not on purpose). If emissions *were* proportional to the stocks, the emission process would be called linear, receptor-controlled.[7]

In our case, $E_N + E_S$ is constant in time and independent of X_N and X_S. Its value can be determined using information in the problem

---

7. The exponential growth of a biological population results from linear, receptor-controlled inflow, as will be discussed below, in Chapter III, Section C.

statement and the Appendix. According to the Appendix (VII.2), natural gas was burned (in 1980) at a rate of 6×10^{19} J/yr, and the energy content of natural gas is 4×10^7 J/m^3 (STP). Because one mole of any gas (STP) occupies 22.4 liters, there are 44.6 moles of gas in a cubic meter. Therefore, natural gas was burned at a rate, R, given by

$$R = \frac{6 \times 10^{19} \text{ J/yr}}{4 \times 10^7 \text{ J/m}^3} \times 44.6 \text{ moles/m}^3 \tag{1}$$

$$= 6.7 \times 10^{13} \text{ moles/yr}.$$

Since natural gas escapes to the atmosphere at a rate equal to 3% of R, and since 6% (on a mole-per-mole basis) of natural gas is ethane (see Appendix [VII.3]),

$$E_N + E_S = (0.03)(0.06)(6.7 \times 10^{13} \text{ moles/yr}) \tag{2}$$

$$= 1.2 \times 10^{11} \text{ moles/yr}.$$

We have determined only the sum of E_N and E_S. However, because nearly all natural gas is mined and vented in the northern hemisphere, we will assume, below, that $E_S = 0$ (but see Exercise 4).

To proceed, we must write the relations among the X's and $E_N + E_S$, which we know, and α and β, which we want to know. To do this we write the steady-state equations, but we should be alert to the possibility that this is misleading. In particular, if $E_N + E_S$ is increasing rapidly enough with time, then X_N and X_S may also be increasing rapidly with time—in which case a steady-state model will be inadequate. Exercise 7 explores this possibility. For now, we ignore it.

The steady-state conditions on X_N and X_S are, respectively:

$$E_N + \alpha X_S = \alpha X_N + \beta X_N \tag{3}$$

and

$$E_S + \alpha X_N = \alpha X_S + \beta X_S. \tag{4}$$

These equations simply state that total inflow to each box (left-hand sides) equals total outflow (right-hand sides). Using this model, we can solve the problem.

If Eqs. 3 and 4 are added together,

$$E_N + E_S + \alpha X_S + \alpha X_N = \alpha X_N + \alpha X_S + \beta X_N + \beta X_S, \tag{5}$$

or

$$E_N + E_S = \beta(X_N + X_S). \tag{6}$$

This can be rewritten as

$$\beta = \frac{E_N + E_S}{X_N + X_S}. \tag{7}$$

Substitution of Eq. 7 into Eq. 3 yields a result for $\alpha(X_N - X_S)$, the interhemispheric flow rate:

$$\alpha(X_N - X_S) = E_N - \frac{(E_N + E_S)X_N}{X_N + X_S} \tag{8}$$

$$= \frac{E_N X_S - E_S X_N}{X_N + X_S}$$

X_N and X_S can be determined from the known values of C_N and C_S by multiplying each concentration by the volume of the hemisphere's atmosphere. This is most easily done on a mole-per-mole basis (see the discussion in Problem II.4). From the Appendix we learn that the number of moles in the atmosphere is 1.8×10^{20}, so

$$X_N = \frac{1.8 \times 10^{20} \text{ moles(air)}}{2 \text{ hemispheres}} \times 1 \times 10^{-9} \frac{\text{moles(ethane)}}{\text{moles(air)}} \tag{9}$$

$$= 0.9 \times 10^{11} \text{ moles(ethane)}.$$

Similarly,

$$X_S = 0.45 \times 10^{11} \text{ moles(ethane)}. \tag{10}$$

Substituting Eqs. 2, 9, and 10 into Eq. 8, and assuming $E_S = 0$, we determine the net rate of flow of ethane across the equator:

$$\alpha(X_N - X_S) = 0.40 \times 10^{11} \text{ moles/yr}. \tag{11}$$

EXERCISE 1: Express $\alpha(X_N - X_S)$ in units of tonnes/yr.

EXERCISE 2: The time constant, α^{-1}, characterizes interhemispheric mixing of air and its constituent trace gases. What is its value? Give an intuitive interpretation of this time constant.

EXERCISE 3: What is the value of the constant β^{-1}, which characterizes removal of ethane from either hemisphere?

EXERCISE 4: If the northern hemisphere were responsible for only 80% of all emissions, then at what rate would ethane flow across the equator and what would α^{-1} be? Given the information provided in the problem statement, how can you be sure that the northern hemisphere is responsible for at least two thirds of the ethane emissions?

EXERCISE 5: Referring to Exercise 2 of Problem II.4, estimate what fraction of all the ethane exiting the troposphere does so by passing to the stratosphere.

EXERCISE 6: Singh et al. (1979) emphasize the possibility that C_2H_6 is removed from the troposphere by chemical reaction with a reactive substance called a hydroxide radical, which is designated by the chemical symbol OH. OH is in such short supply in the atmosphere that the rate at which it reacts with C_2H_6 will depend on how much OH is present rather than on how much C_2H_6 is present. For an analogy, consider the number of beachcombers on a fairly crowded beach who find exotic shells. If the number of such shells washed ashore doubles, then so will the number of findings; but if the beach gets twice as crowded with beachcombers and the number of shells washed ashore remains constant, the number of findings will remain nearly constant. The beachcombers are like the C_2H_6 and the seashells washed ashore are like OH radicals. (*a*) If this C_2H_6 removal process is important, then because OH is likely to be rate-limiting, it is likely that the exit rate is independent of X_N and X_S. Recalculate the value of $\alpha(X_N - X_S)$, assuming both of the exit rates, βX_N and βX_S, are replaced by a common constant, $\beta_N = \beta_S$. Assume that $E_S = 0$. (*b*) The OH concentration in the southern hemisphere may exceed that in the northern hemisphere by about a factor of two because carbon monoxide, which removes OH from the atmosphere, is more abundant in the northern hemisphere's atmosphere. This would result in a greater exit rate in the south than in the north. Assuming $\beta_N = \frac{1}{2}(\beta_S)$ and $E_S = 0$, recalculate $\alpha(X_N - X_S)$.

EXERCISE 7: If $E_N + E_S$ is growing at a rate of 2% per year, explain why the steady-state assumption should still yield a good approximation for α^{-1} and β^{-1}, given the present values of X_N and X_S.

8. A Perturbed Phosphorus Cycle (I)

The box model shown in Figure II-8 can be used to study phosphorus cycling in a lake. In the model, X_1 represents the amount of phosphorus (P) in living biomass, X_2 represents the amount of phosphorus in inorganic form, and X_3 represents the amount of phosphorus in dead organic material. Each X_i is in units of micromoles of phosphorus per liter of lake water. F_{ij} is the flow of phosphorus from stock i to stock j. In the steady state, $X_1 = 0.2$ micromoles(P)/liter, $X_2 = 0.1$ micromoles(P)/liter, $X_3 = 1.0$ micromoles(P) /liter, and the residence time of phosphorus in living biomass is 4 days. Assume that at time $t = 0$, the system is perturbed by the sudden addition of 0.02 micromoles(P)/liter to the inorganic phosphorus compartment, but the rate constants α, β, and γ remain unchanged. When a new steady state is reached, how much phosphorus will be in each compartment?

. . . .

Before calculating numbers, it is helpful to think about the structure of the model. The flows from X_1 to X_3 (resulting from death and excretion) and from X_3 to X_2 (resulting from decomposition) are assumed to be linear, donor-controlled processes. The flow from X_2 to X_1 is both donor- and receptor-controlled. Receptor-control results from the fact that the greater the biomass of living organisms, the greater will be the rate of inorganic phosphorus uptake by that biomass. In contrast, a large pool of inorganic phosphorus, X_2, does not

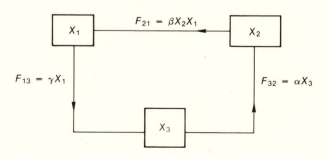

Figure II-8 The inflows and outflows in a three-box model of nutrient cycling.

cause more decomposition and thus does not increase the flow from X_3 to X_2. Similarly, a large pool, X_3, does not increase the death rate and hence does not result in a greater flow from X_1 to X_3. Later on (Chapter III, Problems 10–15) we will discuss biological models that incorporate other assumptions about the dependence of inflows and outflows on compartment variables.

The first step in determining how the system responds to the perturbation of added phosphorus is to determine the numerical values of the rate constants α, β, and γ. Let the initial steady state be characterized by values of the X_i denoted $\overline{X}_i$. The steady-state conditions are derived by setting the inflow to each box equal to the outflow from that same box:

$$\beta \overline{X}_2 \overline{X}_1 = \gamma \overline{X}_1 , \tag{1}$$

$$\alpha \overline{X}_3 = \beta \overline{X}_2 \overline{X}_1 , \tag{2}$$

and

$$\gamma \overline{X}_1 = \alpha \overline{X}_3 . \tag{3}$$

With numerical values for the $\overline{X}_i$ substituted in, we get

$$0.02\beta = 0.2\gamma , \tag{4}$$

$$\alpha = 0.02\beta , \tag{5}$$

and

$$0.2\gamma = \alpha . \tag{6}$$

Eq. 6 can be derived from Eqs. 4 and 5 (just as Eq. 3 can be derived from Eqs. 1 and 2). Hence, Eq. 6 supplies no information[8] not already contained in Eqs. 4 and 5. This redundancy means that Eqs. 4–6 do not provide enough constraints to determine the three rate constants. To determine them we have to use one more piece of information— the fact that the residence time of P in living biomass is 4 days. That residence time is the stock of P in living biomass, $\overline{X}_1$, divided by the flow of P in or out in steady state, $\gamma \overline{X}_1$. So

$$\frac{\overline{X}_1}{\gamma \overline{X}_1} = 4 \text{ days.} \tag{7}$$

---

8. Redundancy of the steady-state equations follows from our assumption in Figure II–8 that the phosphorus flows in a closed cycle (i.e., prior to and also after the perturbation at $t = 0$, no phosphorus flows into any box from outside the system, and no phosphorus leaves the three-box system).

Combining Eq. 7 with Eqs. 4 and 5, it follows that

$$\gamma = (4 \text{ days})^{-1}, \tag{8}$$

$$\beta = (0.4 \text{ days})^{-1}[\text{micromoles(P)/liter}]^{-1}, \tag{9}$$

and

$$\alpha = (20 \text{ days})^{-1}. \tag{10}$$

Now we can solve the problem easily. Call the new values of the X_i, after the perturbation and after a new steady state is reached, $\overline{X}'_i$. The $\overline{X}'_i$ must satisfy the same steady-state equations (Eqs. 1–3) satisfied by the $\overline{X}_i$, because the rate constants have not changed.

For the moment we will omit the cumbersome units of days and micromoles(P)/liter from our notation, but keep them in mind.

The new steady-state conditions are

$$\frac{\overline{X}'_1 \overline{X}'_2}{0.4} = \frac{\overline{X}'_1}{4} \tag{11}$$

and

$$\frac{\overline{X}'_3}{20} = \frac{\overline{X}'_1 \overline{X}'_2}{0.4}. \tag{12}$$

We will not bother writing the third steady-state equation because we now know the information in it can be derived from Eqs. 11 and 12. Eq. 11 tells us

$$\overline{X}'_2 = 0.1 \tag{13}$$

[in units of micromoles(P)/liter] and thus it hasn't changed at all. Eq. 12 then tells us

$$\overline{X}'_3 = 5\overline{X}'_1. \tag{14}$$

Finally, we can make use of the fact that the addition of phosphorus was 0.02 micromoles(P)/liter. Because phosphorus flows in a closed cycle, the total amount present initially plus the amount added to the system at $t = 0$ must equal the total amount present for all times subsequent to $t = 0$. Hence,

$$\begin{aligned}
\overline{X}'_1 + \overline{X}'_2 + \overline{X}'_3 &= 0.02 + \overline{X}_1 + \overline{X}_2 + \overline{X}_3 \\
&= 0.02 + 0.2 + 0.1 + 1 \tag{15} \\
&= 1.320.
\end{aligned}$$

Substituting Eqs. 13 and 14 into Eq. 15,

$$\overline{X}_1' + 0.1 + 5\overline{X}_1' = 1.320, \tag{16}$$

or

$$\overline{X}_1' = \frac{1.220}{6} = 0.203 \text{ micromoles(P)/liter;} \tag{17}$$

and, by Eq. 15,

$$\overline{X}_3' = 1.017 \text{ micromoles(P)/liter.} \tag{18}$$

Note that our answer is independent of where the extra phosphorus was initially placed. Why was it the case that $\overline{X}_i' = \overline{X}_i$ only for $i = 2$? The clue to this is Eq. 1, which tells us that in the steady state, X_2 is a constant equal to γ/β.

EXERCISE 1: Give a qualitative explanation of why the flow from X_2 to X_1 could be assumed to be proportional to the product of X_1 and X_2 while the other flows are simply donor controlled (see Problem II.7)?

* EXERCISE 2: The conversion of phosphorus from organic to inorganic form is carried out by an enzyme called phosphatase. Suppose the initial perturbation at $t = 0$ had been a change in the rate constant governing this conversion. In particular, assume that a permanent 10% reduction in α occurred because a phosphatase inhibitor was dumped into the lake (acid rain could cause such inhibition). When a new steady state is reached, how much phosphorus will be in each compartment?

* EXERCISE 3: Termites process about 13×10^9 tonnes of carbon per year,[9] removing it from dead organic matter and returning it to the atmosphere in the form of CO_2. If Earth's termite populations are destroyed and no other decomposer organisms fill their niche, by how much would the amount of carbon in the atmosphere, in living terrestrial vegetation, and in dead terrestrial organic matter change? You may assume a box model of the same form as the phosphorus model, with X_1, X_2, and X_3 referring to the amounts of carbon in living terrestrial vegetation, the atmosphere, and dead terrestrial organic matter, respectively. Ignore any changes in the amount of carbon in the oceans or in the rate of exchange of carbon between the oceans and the atmosphere. You may also assume that the decrease

---

9. Zimmerman et al. (1982) present the results of some fascinating research on the immense ecological role of termites.

in decomposition activity upon the demise of the termites is reflected in a proportional decrease in the rate constant α. The Appendix (XIII.1,5) provides the information needed to determine numerical values for the $\overline{X}_i$ and for α, β, and γ for the carbon cycle.

9. Where Would All the Water Go?

If evapotranspiration from Earth's land area were to diminish by 20% uniformly over the land area, as might result from widespread removal of vegetation, what changes would occur in the globally averaged precipitation on the land surface and in the globally averaged runoff from the land to the sea?

. . . .

Because we need not consider stocks or time constants, this is not a residence-time problem. It is a box-model problem, however, requiring careful identification of boxes and balancing of inflows and outflows for each box. An immediate guess might be that precipitation would decrease by 20%; but this would be correct only if the hydrocycle did not link our two boxes, the continents and the seas. The existence of runoff from the land to the sea implies that some evaporation from the sea falls as precipitation on the land. Because this portion of land precipitation will not be affected by the 20% decrease in evapotranspiration from the land, the overall effect will be less than 20%.

To solve the problem, a systematic look at the global water budget is helpful. The following water flow rates can be defined:

P_L = rate of precipitation on the land
P_S = rate of precipitation on the sea
R = rate of runoff from the land to the sea
E_{LL} = rate of evapotranspiration from the land that falls as precipitation on the land
E_{LS} = rate of evapotranspiration from the land that falls as precipitation on the sea
E_{SS} = rate of evaporation from the sea that falls as precipitation on the sea
E_{SL} = rate of evaporation from the sea that falls as precipitation on the land

These flows are illustrated in Figure II-9.

Our problem can now be restated in terms of these definitions: How will R and P_L change if E_{LL} and E_{LS} both diminish by 20%? There are three water-conservation relations among the seven quantities we have defined. The first relation states that water is conserved in the sea:

$$P_S + R = E_{SS} + E_{SL}. \qquad (1)$$

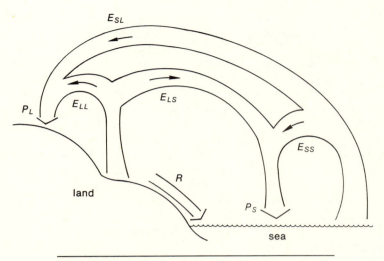

E_{SL}

P_L E_{LL} E_{LS}

E_{SS}

R

land

P_S

sea

Figure II-9 The flows of water in the global hydrocycle.

The second states that water is conserved on land:[10]

$$P_L = R + E_{LL} + E_{LS},\qquad(2)$$

and the third states that the rate of water flow from land to sea equals the rate from sea to land:

$$R + E_{LS} = E_{SL}.\qquad(3)$$

Any two of these can be derived from the third plus the two identities that follow from the definitions:

$$P_L = E_{LL} + E_{SL},\qquad(4)$$

and

$$P_S = E_{SS} + E_{LS}.\qquad(5)$$

All told, there are three independent relations among the seven quantities. Thus, four independent empirical values are needed to determine all seven quantities.

In the Appendix (VI.3), values for P_L, P_S, and R are given. The fourth piece of information we need is that approximately 25% of the evapotranspiration from the land precipitates on the sea, while 75%,

10. This equation is correct only if the stock of groundwater is unchanged. Exercise 2 asks you to examine this assumption.

or three times as much, precipitates on the land ($E_{LL} = 3E_{LS}$). (Exercise 4, below, suggests a procedure for deriving this estimate of the ratio of E_{LL} to E_{LS}. The value of 3 given here is only approximate; a precise empirical value is not available.) With this information and the equations above, we obtain the following values: $P_L = 108 \times 10^3$ km³/yr; $P_S = 410 \times 10^3$ km³/yr; $R = 46 \times 10^3$ km³/yr; $E_{LL} = 46.5 \times 10^3$ km³/yr; $E_{LS} = 15.5 \times 10^3$ km³/yr; $E_{SS} = 394.5 \times 10^3$ km³/yr; and $E_{SL} = 61.5 \times 10^3$ km³/yr. As you may suspect, nonsignificant figures have accumulated here; we can drop them later.

If the reduction in evapotranspiration is uniformly distributed over the land, it is reasonable to assume that E_{LL} and E_{LS} each decrease by 20%. (Note, on the other hand, that if the decrease in evapotranspiration occurred primarily along the coastlines of the continents, then this assumption would be a poor one.) Using primed quantities (P'_L, P'_S, R', etc.) to denote the rates subsequent to the 20% decrease in evapotranspiration, we can write

$$E'_{SS} = E_{SS},$$
$$E'_{SL} = E_{SL}, \tag{6}$$
$$E'_{LL} = .8\, E_{LL},$$

and

$$E'_{LS} = 0.8\, E_{LS}.$$

Then, setting up new conservation equations and identities for the primed quantities, the primed versions of Eq. 3 and 4 become:

$$R' = E'_{SL} - E'_{LS} \tag{3'}$$

and

$$P'_L = E'_{LL} + E'_{SL}. \tag{4'}$$

Use of Eq. 6 then leads to

$$R' = E_{SL} - 0.8\, E_{LS}, \tag{7}$$

$$P'_L = 0.8\, E_{LL} + E_{SL}. \tag{8}$$

Using Eqs. 3 and 4, these can be rewritten as

$$R' = R + 0.2\, E_{LS} \tag{9}$$

and

$$P'_L = P_L - 0.2\, E_{LL}. \tag{10}$$

Numerically, $R' = (46.0 + 3.10) \times 10^3$ km³/yr, which is about a 7% increase over R; and $P'_L = (108 - 9.30) \times 10^3$ km³/yr, which is about a 9% decrease from P_L.

Some readers will be able to solve this problem without plodding through all the formal steps presented here. It is helpful to see how to formalize problems like this one, however, because more complicated situations often make this approach necessary.

EXERCISE 1: By how much will the total global precipitation rate change?

* EXERCISE 2: Changes in hydrologic flows will result in changes in stocks and/or residence times. If the residence times for water in each of the major water compartments on Earth (atmosphere, ocean, ice, soil moisture, groundwater, lakes, and streams) are unchanged as a result of the changes in the flows worked out above, try to deduce which stocks of water will increase, which will decrease, and which will stay the same. For each compartment, would you expect residence time or stock of water to remain more nearly constant subsequent to the changes in flows?

* EXERCISE 3: A reduced rate of precipitation on the land, resulting directly from reduced transpiration, may lead to a reduction in plant growth. On the other hand, reduced transpiration may lead to more moist soil, perhaps enhancing plant growth. Either of these changes in plant growth will alter the transpiration rate again. Such a circular chain of events is called a feedback process. Our solution to the main problem, above, was deficient because it neglected feedback effects. Another feedback process that can result from a decreased evapotranspiration rate is a decreased rate of latent heat transfer from the surface upwards, thus increasing both surface temperature and evaporation rate. Removal of vegetation could induce other climate changes, including an increased CO_2 greenhouse warming and an altered radiation balance caused by changes in the surface albedo (reflectance) of the devegetated land. An increased atmospheric CO_2 concentration would probably further decrease the rate at which plants transpire water.

These effects will be discussed quantitatively in subsequent problems (III.7,8), but for now try to describe the qualitative nature of these feedback effects on P'_L. An amusing wrinkle to consider is that with increased runoff the amount of water in the sea would rise, thus increasing its surface area. Would this significantly increase the rate of evaporation from the sea and thus alter our answer?

* EXERCISE 4: Using (a) the residence time for atmospheric water derived in Problem II.2 and (b) a map of the world, estimate the ratio of E_{LL} to E_{LS}.

EXERCISE 5: Deforestation of mountain slopes has been proposed as a way to provide more runoff water for irrigation in the western United States.[11] The idea is that more snowmelt will flow into reservoirs while less will be transpired away by trees. Is this a sensible idea?

---

11. See, for example, the collection of articles on this topic in *The Water Resources Bulletin*, Volume 19, June, 1983.

10. Aluminum in the Himalaya

In a remote area in Nepal, the concentration of aluminum (Al) in outdoor air at ground level averages 9.4×10^{-8} μg/cm^3. (It is much higher inside the Sherpa dwellings because of wood and yak dung burning).[12] At the same site, the Al concentration in the top 1 cm of fresh snow averages 0.12 μg/g, while in the top 1 cm of three-day-old snow it averages 0.20 μg/g. (*a*) Calculate the average deposition velocity of the Al falling to the ground when it is not snowing. (*b*) How large are the particles to which the falling aluminum is attached?

· · · · · · ·

(*a*) This is a variation on the usual stock-flow–residence time problem. You are given a stock, expressed as the concentration of aluminum in the air. A flow of aluminum groundward can be derived from the information about old snow and new snow if you assume that the extra aluminum on the old snow surface is due to deposition. Now, however, you are asked to calculate a deposition velocity rather than a residence time. The reason a deposition velocity and not a residence time can be calculated from the given data is that the stock of aluminum in the atmosphere is expressed in units of mass per unit volume (μg/cm^3) while the flow (see below) must be calculated in units of mass per unit time per unit *area*. Dividing the flow by the stock gives a quantity expressed in units of length per unit time. This quantity is not an inverse residence time but, rather, a deposition velocity.

The flow can be calculated as $0.20 - 0.12 = 0.08$ μg of Al per gram of snow per 3 days (see Figure II-10). This is the rate of increase of aluminum concentration in the top centimeter of snow. Since fairly fresh snow has a density of about 0.1 g/cm^3 (about 0.1 that of water), the aluminum was measured in 0.1 g(snow)/cm^2 of surface. Hence, the flow can be expressed as

$$F = \frac{0.08 \, \frac{\mu g(Al)}{g(snow)} \times 0.1 \, \frac{g(snow)}{cm^2}}{3 \text{ days}}$$

$$= 0.0027 \; \mu g(Al)/cm^2 \text{ day}.$$

(1)

---

12. The data and the inspiration for this problem come from Davidson et al. (1981).

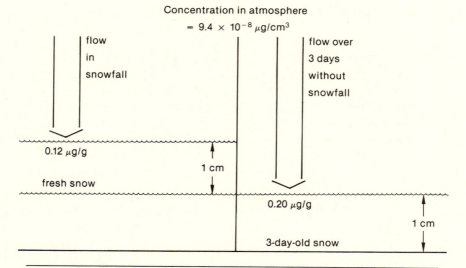

Concentration in atmosphere
$= 9.4 \times 10^{-8}\ \mu g/cm^3$

Figure II-10 A pictorial summary of the input data given in the problem statement.

The stock in the atmosphere is

$$M = 9.4 \times 10^{-8}\ \mu g(Al)/cm^3. \tag{2}$$

Hence, the deposition velocity, F/M, is

$$\frac{F}{M} = 2.9 \times 10^4\ \text{cm/day}$$

$$= 0.34\ \text{cm/sec}. \tag{3}$$

(b) To calculate the size of the particles falling at this speed, a digression is needed. Stokes Law describes the rate at which objects fall through a medium like air or water, provided the velocity of the object is small enough to create no turbulence. The law states that the frictional drag force on a spherical object is

$$F = 6\pi\eta vr, \tag{4}$$

where η is the viscosity of the medium (given in the Appendix V.1), v is the velocity, and r is the object's radius.[13] Stokes Law is applicable provided a certain quantity called the Reynolds number, R, defined by

$$R = \rho_m vr/\eta, \tag{5}$$

---

13. For a nonspherical object, r is replaced by an effective radius. Consult a physics text for details.

is less than about 0.5. Here, ρ_m is the density of the medium; in this case, the medium is air, so $\rho_m \approx 0.001$ g/cm^3. A falling object soon reaches a terminal velocity in which the frictional force retarding its motion equals the gravitational-minus-buoyancy force pulling it downward. For a particle falling through air the buoyancy is negligible and so the downward force is the product of the mass, m, times the acceleration of gravity, g. Hence the terminal velocity can be calculated by setting

$$6\pi\eta vr = mg. \tag{6}$$

This balance of forces is illustrated in Figure II-11.

We know the values of everything in Eq. 6 except r and m. Assuming the particles to which the aluminum is attached are spherical, r is their average radius; m is their average mass. It is a reasonable assumption that the density of the particles, ρ_p, is very roughly equal to 1 g/cm^3, the density of water. This is because small particles falling from the atmosphere are often actually aerosols. Aerosols may contain water; but even if dry, they are comprised of fluffy solids—irregular hollow structures less dense than typical solid materials forming Earth's crust. Thus, we find

$$m = \frac{4}{3}\pi r^3 \rho_p. \tag{7}$$

Rewriting Eq. 6 yields

$$6\pi\eta vr = \frac{4}{3}\pi r^3 \rho_p g \tag{8}$$

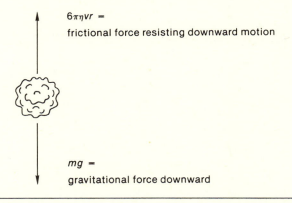

$6\pi\eta vr =$
frictional force resisting downward motion

$mg =$
gravitational force downward

Figure II-11 The forces acting on a small particle in the atmosphere; v, r, and m are the particle's velocity, radius, and mass, respectively, η is the viscosity of air, and g is the acceleration of gravity.

or

$$r = \left(\frac{4.5\eta v}{\rho_p g}\right)^{1/2}. \tag{9}$$

Using Eq. 3 and the values of η and g from the Appendix (V.1), we calculate

$$r = 0.0005 \text{ cm} = 5 \text{ microns.} \tag{10}$$

EXERCISE 1:   Is the required condition on the Reynolds number satisfied in our case?

EXERCISE 2:   *(a)* If all the falling aluminum originated with local biomass fuel combustion, and if the aluminum composition of biomass is taken from the Appendix (VIII), how much fuel per unit area would have to be burned in this region of Nepal in order to account for the observations? *(b)* Making a reasonable assumption about per capita fuel consumption in Nepal, what population density does this correspond to?

EXERCISE 3:   What assumptions were made in deriving a flow of 0.0027 μg(Al)/cm^2-day? Estimate, qualitatively, for each assumption listed, what effect relaxing that assumption would have on the final result.

11. An Indoor Risk

Radioactive radon gas (Rn^{222}) enters an average building at the rate of one picocurie per second per square meter of foundation area. Consider a house with a foundation area of 200 m^2 and an air volume of 1000 m^3. Assume that the house is well designed for energy conservation so that the ventilation rate is low and only one tenth of the air in the house is exchanged with outdoor air every hour. (*a*) What will be the average steady-state concentration of Rn^{222} in the house? (*b*) In the steady state, what whole-body radiation dose, in rads/yr, will an adult male receive directly from Rn^{222} decay if he spends 12 hr a day in the house and derives the entire dose from the decay of radon in the inhaled air in his lungs?

· · · · · · ·

First a note on units and radioactive decay is in order.[14] Every radioactive isotope has a characteristic time constant, called the half-life. It is usually denoted by the symbol $T_{1/2}$. Suppose that at some time, $t = 0$, there are N atoms of an isotope in a closed box. After a period of length $T_{1/2}$ has elapsed, there will be $N/2$ atoms left. The other half will have decayed to form some other atomic species. At time $t = 2T_{1/2}$, $N/4$ will remain, and so on. In each time interval equal to one half-life, half of the radioactive atoms of that isotopic species present at the start of the interval will decay. The Appendix (XI.1) provides values of $T_{1/2}$ for a number of important radioactive isotopes.

Consider a collection of atoms of a particular isotope with half-life $T_{1/2}$. Suppose at time t there are $N(t)$ atoms present. The rate of decay of these atoms is proportional to N itself. If twice as many atoms of the isotope are present then twice as many will decay per unit time. In fact, this is the feature of radioactive decay that makes it possible to assign a constant half-life to each isotope, regardless of the number of atoms of the isotope present. The rate of decay is called the activity and it can be written

$$\text{activity}(t) = \lambda N(t). \tag{1}$$

---

14.   For an introduction to the science of radioactivity and radiation in an environmental context, see Harte and Socolow (1971), Eisenbud (1963), and Ehrlich et al. (1977).

Here, λ is a rate constant related to the half-life by[15]

$$\lambda = \frac{\ln 2}{T_{1/2}} = \frac{0.693}{T_{1/2}}. \tag{2}$$

Note that activity is expressed in units of number (of decaying atoms) per unit time. The curie (Ci) is a unit of activity, corresponding to 3.7×10^{10} decays per second. A picocurie (pCi) is 10^{-12} curie or 0.037 decays per second (see Sections I.1 and I.14 of the Appendix). If, for example, at $t = 0$ there are $N = 10^5$ atoms of an isotope with $T_{1/2} = 20$ sec, then the activity at time $t = 0$ is $(0.693/20) \times 10^5$ or 3,465 decays/sec. To convert this to units of curies, we divide by 3.7×10^{10}; to convert to units of pCi, we divide by 0.037. Thus, there would be 9.4×10^4 pCi of activity in that particular sample. Twenty seconds later, the number of atoms of the isotope that remain will be reduced in half and the activity would then also be reduced by a factor of 2.

(a) Returning to our problem, we must choose sensible units to measure the stock of Rn^{222} in the indoor air. Any of several units—picocuries, molecules, grams, or moles—could be used to describe our variable. Or we could divide any of these by cubic meters or by kilograms of air to obtain a variable expressed in units of concentration.

It is conceptually easiest to work with the number, N, of atoms of Rn^{222} in the house. Converting back and forth between N and activity is easily done using Eq. 1. Similarly, converting either N or activity to units of household concentration is straightforward.

The equilibrium value of N is determined by setting the rate of inflow of Rn^{222} atoms equal to the rate of outflow (see Figure II-12). The rate of inflow of Rn^{222} is determined by the source term, 1 pCi/m²sec. For a house with a foundation area of 200 m², this is 200 pCi/sec or $200 \times 0.037 = 7.4$ decays/sec². That is, Rn^{222} activity enters the house at a rate of 7.4 decays per second per second. We denote by F_{in} the rate at which atoms of Rn^{222} enter the house. Eq. 1 tells us that the number of atoms of an isotope equals λ^{-1} times the activity of that isotope. Therefore, F_{in} equals λ^{-1} times the rate at which activity due to Rn^{222} enters the house. Using the value of $T_{1/2}$ for Rn^{222} (see Section XI.1 of the Appendix) of 3.8 days, or 3.3×10^5 sec, we obtain a rate constant, λ, for Rn^{222} of $0.693/(3.3 \times 10^5$ sec) or 2.1×10^{-6}/sec. Therefore, we determine that

$$F_{in} = \frac{7.4}{\lambda} = 3.5 \times 10^6 \text{ atoms/sec.} \tag{3}$$

---

15.   The reason for this relation will be explained in Problem III.3. You may be able to see where it comes from by referring back to Problem I.5.

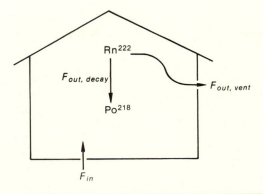

Figure II-12 The inflow and outflow of Rn^{222} for a house. There are two outflow terms: $F_{out,\ decay}$ describes the loss of Rn^{222} by decay to its first daughter, Po^{218}; and $F_{out,\ vent}$ describes the loss of Rn^{222} with the airflow in the house as indoor and outdoor air are exchanged. F_{in} is the source term.

The outflow rate consists of two terms. The first describes loss of Rn^{222} by radioactive decay in the house. This decay rate is simply the activity $\lambda N = 2.1 \times 10^{-6}$ N/sec. The second describes outflow resulting from ventilation. The ventilation rate is 0.1 air exchanges per hour. If the Rn^{222} is well mixed throughout the house, then one tenth of the Rn^{222} in the house is removed every hour as well. Hence, the second term in the outflow rate is 0.1 N/hr or 2.8×10^{-5} N/sec. Combining the two outflow terms, we find

$$
\begin{aligned}
F_{out} &= F_{out,decay} + F_{out,vent} \\
&= (2.1 \times 10^{-6} + 2.8 \times 10^{-5})\,N \\
&= 3.0 \times 10^{-5}\ N/\text{sec.}
\end{aligned}
\tag{4}
$$

The steady-state value of N can now be calculated by equating F_{in} and F_{out}:

$$
3.5 \times 10^{6}/\text{sec} = 3.0 \times 10^{-5}\ N/\text{sec.}
\tag{5}
$$

Hence,

$$
N = 1.17 \times 10^{11}\ \text{atoms.}
\tag{6}
$$

This is the number of atoms of Rn^{222} present in the house in the steady state. The steady-state concentration, C, will equal N/V, where V is the volume of air in the house. Hence,

$$
C = 1.17 \times 10^{11}/1000 = 1.17 \times 10^{8}\ \text{atoms/m}^{3}.
\tag{7}
$$

Multiplying by λ, the steady-state activity concentration is $2.1 \times 10^{-6} \times 1.17 \times 10^8$, or 246 decays/m^3-sec. Using the definition of the curie, this is equivalent to 6.6×10^{-9} Ci/m^3.

(b) The dose of radiation received in the lung of an occupant of the house depends upon the activity of the isotope in the lung. This, in turn, depends on the average number of Rn^{222} atoms in the lung, which is in turn determined by the concentration of Rn^{222} atoms in the indoor air. Section XV of the Appendix indicates that a typical adult male inhales 7.5 liter/min while resting and 20 liters/min while engaged in light activity. The breathing rate at rest is about 7 breaths/min, and so each breath contains about 1 liter of air. During light activity, a person takes about 13 breaths/min and so inhales about 1.5 liters of air with each breath.

Let us then assume that during the 12 hr indoors, the adult male typically has at any time about 1.2 liters of freshly breathed air in his lungs. Eq. 7 tells us that this much air will contain $1.2 \times 10^{-3} \times 1.17 \times 10^8$ or 1.4×10^5 atoms of Rn^{222}. Ignoring old air in the lungs, the activity of the lung air is $\lambda N = 2.1 \times 10^{-6} \times 1.4 \times 10^5$ or 0.29 decays/sec. Thus during a typical year of 12-hr days in the home, our subject's lungs will endure $365 \times 12 \times 3600 \times 0.29$, or 4.6×10^6 Rn^{222} decays.

A rad of radiation dose corresponds to 100 ergs deposited per gram of tissue. To estimate the whole-body dose, we must determine the number of ergs deposited in the body and divide by the number of grams of body weight. The Appendix (XI.1) tells us that each Rn^{222} decay produces an α-particle with 5.5 MeV of energy. One MeV is equivalent to 1.6×10^{-6} ergs, so 4.6×10^6 decays/yr deposits $4.6 \times 10^6 \times 5.5 \times 1.6 \times 10^{-6}$ or 40.5 ergs/yr. A typical adult male has a mass of 7×10^4 g, so the whole-body dose, in rads, is $40.5/(100 \times 7 \times 10^4)$ or 5.8×10^{-6} rads/yr.

The whole-body radiation dose received from all sources by an average person is about 0.2 rads/yr (see Section XI.2 of the Appendix). In comparison, the indoor whole-body exposure from Rn^{222} calculated above is extremely small. However, the indoor Rn^{222} dose is somewhat unusual. The isotope is a gas that emits α-particles. The dose we calculated is deposited entirely in the lungs, where α-particles are particularly dangerous.

The source term used here, 1 pCi/m^{2}sec, is an average over a very wide range of conditions. Rn^{222} is produced from the decay of U^{238}, which produces a series of intermediate isotopes in a decay chain that leads to Rn^{222}. Because the ground distribution of U^{238} is heterogeneous, so is the rate at which Rn^{222} enters buildings. Moreover, the rate at which Rn^{222} penetrates building materials depends on the nature of those materials and the type of construction. Source rates as low as 0.1 and as high as 100 pCi/m^2-sec have been observed. The effect of choosing a different source rate or a different ventilation rate can easily be analyzed with the methods presented here. (see Exercise 3, following)

A more serious difficulty with our approach to estimating the radiation dose from Rn^{222} is that the major dose will result not directly from Rn^{222} decay but from decay of its daughter isotopes. (The decay sequence is provided in the Appendix, (XI.1)) Po^{214}, an α-emitter near the end of the Rn^{222} daughter sequence with a 1.6 msec half-life, poses this problem's greatest health risk. The dose from the daughters of Rn^{222} is difficult to estimate because it depends significantly on their distribution within the house. While Rn^{222}, a gas, tends to mix uniformly throughout indoor air, the daughters do not. They are reactive and can attach to particles. By settling to the floor on large particles or attaching to walls, they escape being vented out of the house and may also escape being inhaled. On the other hand, if the daughters attach to particles of submicron dimensions, they may be breathed deeply into the lung and attach there, thus increasing the dose. All in all, reliable estimates of realistic daughter concentrations and of total doses to indoor occupants are difficult to make and require specification of more information (particularly indoor particle concentrations) than we have provided.[16] Exercise 4 looks at one limiting case.

The issue of regulatory standards for indoor Rn^{222} concentrations is currently under debate. Discussion is focused on the appropriateness of a permissible level of 3×10^{-9} Ci/m^3. In addition, energy conservationists, aware of the potential hazards of indoor Rn^{222}, have become interested in heat-exchange systems for buildings that would permit air ventilation without loss of indoor heat.

EXERCISE 1: In working out our solution to the problem, what did we implicitly assume about the concentration of Rn^{222} in the air outside the house? Give a qualitative argument why that approximation ought to be a good one.

EXERCISE 2: Our dose estimate did not depend on how long air remains in the lungs but only on the average amount of air that is in the lungs at any one time. Why was that the case? Under what circumstances would that independence from time not hold true?

EXERCISE 3: Consider two other houses with the same dimensions as the one here: *(a)* One has a higher ventilation rate of one air exchange per hour; *(b)* the other permits no air exchange at all, so that the only exit pathway for Rn^{222} is radioactive decay. Calculate the steady-state concentration of Rn^{222} activity (in Ci/m^3) in these houses. Although a condition of no air exchange is unrealistic (a rate of 0.1 air exchanges per hour requires very airtight house construction and avoidance of unnecessary open-door time by the occupants), it provides a limiting case for which the Rn^{222} exposure will be maximum.

---

16. For an excellent review of the health risks associated with indoor Rn^{222}, see the entire issue of *Health Physics* 45, No. 2, particularly the article by Nero (1983).

EXERCISE 4: The first daughter of Rn^{222} (i. e., its immediate decay product) is Po^{218}, with a three-minute half-life. Being the decay product of Rn^{222}, Po^{218} is formed at precisely the rate of Rn^{222} decay. In the house with a 0.1/hr ventilation rate, what will be the steady-state quantity of Po^{218}, expressed first in units of number of atoms and then in units of curies, under the assumption that none of the Po^{218} is vented from the house? Is it a numerical accident that the number of curies of Po^{218} equals the number of curies of Rn^{222}?

* *EXERCISE 5:* If, on the average, the emission of Rn^{222} to the atmosphere from the continents is 1 pCi/m^2-sec and from the oceans is zero, estimate the average concentration of Rn^{222} in the atmosphere. Qualitatively, how will the vertical distribution of Rn^{222} compare with that of N_2O? (See Exercise 2 of Problem II.4.)

B.
Thermodynamics and Energy Transfer

Thermodynamics is primarily the study of energy and its interaction with matter. Energy can take many forms: the energy an object has by virtue of its motion is called kinetic energy; the kinetic energy of molecules moving randomly is called heat energy; the energy required to lift or shove an object that is acted on by such forces as gravity and friction is called work; the energy available for release by dropping an elevated object is called potential energy; the energy available for release by burning substances such as coal is called chemical energy; and chemical energy, when released, often takes the form of heat energy.

The ways energy can interact with matter are likewise numerous. Energy can warm matter, melt it, freeze it, boil it, expand or contract it, scramble or unscramble it, or mutate it. The results depend on what kinds of energy and matter are involved. Despite the apparent complexity, however, a few simple and universal physical laws have been discovered that allow us to make sense of the whole story. Among these are the laws of thermodynamics. The first law states that the total amount of energy in a closed system (i.e., one which energy cannot enter or leave) remains constant, regardless of the transformations among energy types within the system. This law plays a central role in modeling both meteorological phenomena and physiological aspects of animal interactions with their environment.[17] The second law, which describes constraints on the conversion of one type of energy to another, is particularly useful for evaluating the

---

17. Schmidt-Nielsen (1972) has written a delightful and informative book on animal physiology and physics. An excellent introduction to the subject of thermodynamics is the terse text by Van Ness (1969). *Entropy for Biologists* by Morowitz (1970) is also highly recommended.

performance of technological devices designed to convert one form of energy to another. The first six problems that follow illustrate various applications of thermodynamics. The last problem in this section, on the multiple scattering of solar radiation in the atmosphere, illustrates concepts about energy transfer in the atmosphere that are widely used in climate research.

12. Electricity from Junk Mail

At about what rate could electricity be produced by burning everybody's junk mail in the United States, assuming a conversion efficiency from heat to electricity of 30%?

· · · · · · ·

Three sequential questions lurk in this problem: How much junk mail would be burned, how much heat would be produced from that fuel, and how much electricity would be produced from that heat? We tackle each in turn.

In the United States we number about 7×10^7 households. Let's assume each receives as much junk mail as mine does. After weighing the ads and pleas in my mailbox for two weeks, I estimate that I am delivered about 2.0×10^4 g of paper junk mail per year. Hence, in the United States, about $2.0 \times 10^4 \times 7 \times 10^7$ or 1.4×10^{12} g/yr of paper are available for conversion to electricity.

According to Section VII.4 of the Appendix, burning paper produces about 2×10^4 J/g. U.S. junk mail would thus produce heat at an annual rate, R, of about

$$R = 1.4 \times 10^{12} \text{ g/yr} \times 2 \times 10^4 \text{ J/g} \tag{1}$$
$$= 2.8 \times 10^{16} \text{ J/yr.}$$

How much electric power would this produce? The problem states that the conversion of heat to electricity proceeds at 30% efficiency.[18] This means that

rate of electricity production = (0.3)(rate of heat production).   (2)

Electricity would be produced at a rate, R', given by

$$R' = (0.3)(2.8 \times 10^{16} \text{ J/yr}) \tag{3}$$
$$= 8.4 \times 10^{15} \text{ J/yr.}$$

A more familiar unit for the rate of electricity production (or conservation) is the watt (W), a unit of power. Power is the rate of energy flow, and a watt is defined as the flow of one joule per second. If you conserve a watt of power for a period of one second, you have conserved one joule of energy. Since there are 3.15×10^7 seconds

---

18. Problems II.15 and 16 delve more deeply into the subject of conversion efficiency and explain why heat-to-electricity conversion efficiency cannot be 100%.

in a year, electric energy would be produced at a rate of $8.4 \times 10^{15}/3.15 \times 10^7 = 2.67 \times 10^8$ W. This is roughly one fourth the power output from one typical, modern, nuclear or coal-fired electric power plant.

EXERCISE 1:   What fraction of the total rate of U.S. electricity production in 1980 does the answer above correspond to?

EXERCISE 2:   What percentage of total world net primary productivity would have to be harvested and burned, at 30% conversion efficiency, to meet all the world's electricity needs, at the rate of electricity consumption in 1980?

* EXERCISE 3:   In addition to his many discoveries in pure science and his useful inventions, Benjamin Franklin also was responsible for numerous social innovations. Indeed, junk mail is delivered by a postal system that was initiated in the American colonies by Franklin. He also was a pioneer in what is now called energy-impact analysis: Specifically, in his wise and jocular manner, Franklin proposed that Paris switch from standard time to daylight savings time, based on his estimate that such a scheme would save candle wax (Goodman 1956). The next time your area switches to or from daylight savings time, keep track of the amount of electric energy you use for lighting each day in your home during the week before and after the switch. (The electric energy used to light a *W*-watt bulb for *S* seconds is *WS* joules.)

13. How Hot Is Planet Earth?

What is the temperature of planet Earth? This is a complex question because Earth's temperature varies with time and especially with location. So, to be more specific, imagine a space traveler viewing our planet from afar. What temperature would she say Earth is?

. . . .

At the level of analysis in this problem, a precise answer to the question "How hot is Earth?" is derived. However, the answer will not be of direct relevance to our experience on Earth's surface. In later problems (III.6–9), more realistic models will be used to calculate more pertinent information, such as Earth's average surface temperature. These models will build upon the simple "core model" presented here.

Earth, continually receiving energy from the sun, would get hotter and hotter if it did not radiate energy back to space. This insight suggests that an energy conservation equation, balancing the rate of energy flow arriving at Earth against the rate of energy flow departing, might allow a calculation of the earth's temperature. To carry out this balancing act, the input and output rates are needed. Because only inputs and outputs are involved at this stage, you can ignore the complexities of atmospheric convection and the hydrocycle, which serve to redistribute energy on the wet, airy, planetary skin but do not affect the overall energy balance of the planet. Only radiant energy can be transmitted to any appreciable degree through the nearly empty space beyond Earth's atmosphere.

Consider, first, the solar input. On a flat disc perpendicular to the unimpeded sun's rays and 150×10^6 km from the sun, where Earth resides, the solar flux is 1,372 W/m². For future reference, this important number will be denoted by the symbol, Ω. Across an average square meter above the atmosphere the flow of solar energy will not be this large. On the night side of Earth the flow will obviously be zero, and only on the point of Earth directly under the sun (for example, the equator at high noon on the equinox) will the energy flux from the sun equal Ω. The flux above the atmosphere, averaged over day and night and over all latitudes, is equal to $\Omega/4$. To see why this is so, note, first, that the ratio of the area of a sphere to the area of a disc of the same radius is $4\pi\, r^2/\pi r^2 = 4$. Thus, if you imagine the sun's rays making dots on the imaginary disc as they cross it, and then stretch the disc around the sphere, the dots will be one fourth as densely packed (see Figure II-13).

We must now determine the rate of outgoing energy, per unit area of the Earth's surface. If you looked at the earth from space, you

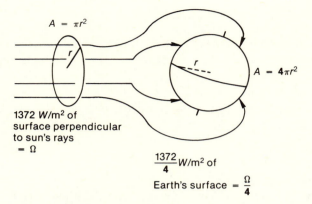

Figure II-13 The solar flux on Earth. The factor of four arises because the surface area of Earth is four times the area of a disc with a diameter equal to that of Earth. The solar flux on Earth, averaged over all latitudes and longitudes, is $\Omega/4$. The sun's rays do not of course, strike the Earth from behind. Rather, Earth turns and the radiation is distributed over all longitudes.

would see a moon-like glow on its sunlit side. This is reflected solar radiation. Earth has a reflection coefficient, called the albedo, of about 0.30, which means that about 30% of the incident solar flux is reflected directly back to outer space. (Most of this radiation is reflected by Earth's clouds rather than the planet's surface.) The remaining 70% of the solar flux is absorbed, either in the atmosphere or by materials at Earth's surface. This absorbed solar energy is converted to infrared energy by the absorbing material. It is then emitted from Earth and radiated away to space in a form called "infrared radiation."

Thus, the outgoing energy takes two forms: reflected solar energy, and infrared radiation by objects that have absorbed such energy. The rate at which an object radiates energy away from itself depends on its temperature; the hotter the object, the greater the rate of radiation. Objects that are maximally effective radiators of energy are called blackbodies. (A glowing lump of coal is a good example.) Such objects are also perfect absorbers, which is why they are called black. The Stefan-Boltzmann Law describes the relation between the temperature of such objects and their rate of radiation. It states:

$$F = \sigma T^4. \tag{1}$$

Here, F is the radiant energy output, with units of energy per unit time per unit area, and T is the absolute temperature of the object in kelvins. The fundamental constant, σ, is given by

$$\sigma = 5.67 \times 10^{-8} \text{ J/m}^2 \text{ sec K}^4 \tag{2}$$

To a good approximation, Earth is a perfect emitter and absorber of infrared radiation, and so Eqs. 1 and 2 can be used to describe the outgoing infrared radiation.[19] Thus, as shown in Figure II-14, the incoming and outgoing energy fluxes are:

$$F_{in} = \Omega/4$$

$$F_{out} = a\Omega/4 + \sigma T^4,$$

(3)

where the albedo is denoted by a. Equating incoming flux to outgoing flux,

$$\Omega/4 = a\Omega/4 + \sigma T^4.$$

(4)

Eq. 4 implies

$$T^4 = \frac{(1 - a)\Omega}{4\sigma}.$$

(5)

The solution to this equation is often denoted by T_0, a quantity referred to as the blackbody temperature of Earth. Using the numerical values of Ω, a, and σ, the result is $T_0 = 255$ K.

T_0 is not equal to the measured average temperature of Earth's surface, which is about 290 K. Earth's surface does emit infrared radiation upward, but that radiation is largely absorbed in the atmosphere

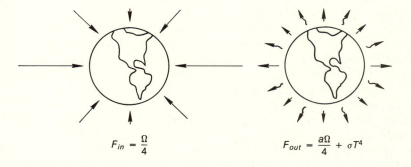

$$F_{in} = \frac{\Omega}{4} \qquad\qquad F_{out} = \frac{a\Omega}{4} + \sigma T^4$$

Figure II-14 The inflow and outflow of energy on Earth. Straight arrows are solar flux and wiggly ones are infrared radiation. Arrow lengths are roughly proportional to actual flows. Note that the total outflow from the polar regions is greater than the inflow, whereas the outflow from the tropics is less than the inflow. Poleward flow of heat in the oceans and atmosphere on Earth accounts for this.

---

19. More generally, any object radiates at a rate $\epsilon\sigma T^4$ where ϵ is a function of the wavelength of the radiation and is always ≤ 1. An object is a blackbody in that portion of the spectrum of wavelengths for which $\epsilon = 1$.

and then reradiated within the atmosphere (see Problem III.6). Radiation emitted higher in the atmosphere has a greater chance of being radiated directly to outer space. The zone of the atmosphere that radiates directly to outer space consists of air from the mid-troposphere upward, and the density-weighted mean temperature of this zone is about 250 K, which is quite close to our calculated value of 255 K. Observed from far away, the infrared earthglow has a spectrum roughly like that of a 250-K blackbody, and thus our assumption that this planet is thermodynamically "black" is not a bad one. It must be emphasized, however, that Earth is nearly black only in the infrared range of the spectrum, not in the spectral range of sunlight. A planet that was a blackbody in the solar spectrum would have zero albedo because it would be a perfect absorber of sunlight; in contrast, Earth's albedo in sunlight is about 30%.

EXERCISE 1: Verify that Eq. 5 yields the result: $T_0 = 255$ K.

EXERCISE 2: Using the procedures presented here, we calculate an infrared blackbody temperature for Jupiter of 98 K; the measured value is 130 K. It is the only planet with such a large discrepancy. The reason is that the gravitational collapse of Jupiter's interior produces a sizeable amount of internal heat energy. Estimate the internal energy flux on Jupiter from the data above and compare it to Earth's geothermal flux, which is determined from data given in the Appendix (VII.1).

EXERCISE 3: Throughout Earth's history there have been cycles and fluctuations in the planet's average temperature. Changes in Earth's solar orbit have been proposed as one cause of these variations.[20] While the most important orbital modification of Earth's climate is most likely due to changes in the orientation of the tilt of Earth's axis with respect to orbital perihelion and aphelion (an effect too complicated to study with the globally averaged model presented here), a simpler phenomenon—change in the mean distance between Earth and the sun—can be studied. Assuming a circular orbit, by how much would the distance from Earth to the sun have to change in order to bring about a 1 K increase in T_0? Later, in the second section of Chapter III, other possible causes of climate alteration will be discussed.

EXERCISE 4: Derive the fact that the sun's blackbody temperature is 5,750 K, given the values of Ω and σ quoted above, and the information from the Appendix (IV) on the distance from the sun to Earth and the radius of the sun.

---

20. See Ruddiman and McIntyre (1981) and Kerr (1983a) for further discussion of this hypothesis.

* *EXERCISE 5:*   Imagine a very accurate, high-resolution light meter capable of measuring the light intensity reaching Earth's surface at night from the bright half of a half moon and, separately, the very faint light intensity from the "dark" half. How could you calculate Earth's albedo, knowing these two light intensities and some relevant distances?

14. Milk and Muscle

How high can you climb on a liter of milk?

. . . .

The chemical energy contained in the food you eat serves many functions. Some of it is used to maintain body temperature and drive bodily metabolic processes. Some is used to produce new biomass, either for net growth or to replace worn-out tissue. Some is used to perform mechanical work such as climbing a hill or pushing a wheelbarrow. And, of course, some of the chemical energy is excreted, playing a further ecological role but no longer available directly to your body. These energy expenditures are linked in a number of interesting ways. For instance, the performance of mechanical work generates internal heat; new biomass cannot be produced without the aid of energy-consuming metabolic processes; and much muscular activity performs internal work to drive circulation, digestion, and excretion.

What fraction of the energy in the food we eat can be converted to muscular work? The following qualitative argument suggests that the fraction of the food energy converted to mechanical work is small. Suppose you engage in mild exercise or physical work after a hearty breakfast. Chances are you will be hungry again in three or four hours. If you drive an automobile for three or four hours after the same breakfast, you will probably be about equally hungry. A sedentary activity burns food about as fast as a muscular one, and thus a large fraction of our caloric intake is used for "maintenance" and is not available for muscular work. Physiologists tell us that rarely is more than 25% of our caloric intake convertible to muscular work (see Exercise 4 below). To solve our present problem, I will assume a 15% conversion. The important thing is to learn how to set up the problem; if you find a better conversion rate you can redo the calculations later.

A liter of milk has an energy content of about 2.4×10^6 J (see Appendix, VII.4). To lift a mass, m, to a height, h, against the acceleration of gravity, g, on Earth requires an amount of work, W, given by

$$W = mgh. \tag{1}$$

With a conversion efficiency labeled ϵ, the amount of work that can be performed, W, is related to the food energy input, Q, by

$$W = \epsilon Q. \tag{2}$$

Assuming a mass of 50 kg, and a conversion efficiency of 15%, and using $g = 9.8$ m/sec^2, the result is

$$h = \text{(conversion efficiency)} (Q)/mg$$

$$= \frac{(0.15)(2.4 \times 10^6)\,(kg\ m^2/sec^2)}{(50\ kg)\,(9.8\ m/sec^2)} = 700\ m. \tag{3}$$

This is a plausible result. After an average meal (which has about 1.5 times the calorie content of a liter of milk) if you quickly climb only a few hundred vertical meters you will not feel hungry right away, whereas if you climb a thousand vertical meters in three or four hours you will feel as if you have *more* than burned up that last meal!

EXERCISE 1: (a) Express the rate at which a typical adult human being produces body heat, in units of watts, assuming 80% of the ingested calories are metabolized. (b) Using Table XV in the Appendix, calculate the resting metabolic rate of an average adult human being.

EXERCISE 2: A typical adult Alaskan brown bear (*Ursus arctos*) will enter "hibernation" weighing about 450 kg and emerge six months later having burned up about 100 kg of fat. Using Table VII in the Appendix, estimate the wintertime metabolic rate of the bear.

EXERCISE 3: It has been observed (see Schmidt-Nielsen 1972 for a review) that the resting metabolic rate of animals is proportional to the three-fourths power of body weight. Assuming that scaling relation and using the resting metabolic rate of human beings from Exercise 1, estimate the resting metabolic rate of the bear and compare that value with the metabolic rate during "hibernation" from Exercise 2. True hibernators, such as rodents, will metabolize during hibernation at a rate well below half their normal resting metabolic rate. Is the Alaskan brown bear a true hibernator?

* *EXERCISE 4:* The human body gets rid of heat produced by metabolic activity via radiation, evaporation of water (perspiration), and conduction to the surrounding air. Knowing the rate at which water is evaporated, we can estimate the rate of cooling by that process, as discussed later in Problem II.16. The Stefan-Boltzmann Law (see Problem II.13) can be used now, however, to describe radiation from a blackbody at temperature 310 K (98.6° F) to a surrounding environment. Hence, a lower limit on sedentary heat losses is obtained. It is a lower limit because only one mechanism of heat removal (radiation) is estimated. With a lower limit on heat loss, we obtain an upper limit on the fraction of our daily caloric intake available for mechanical work. Doing so is not easy, but the ambitious reader may want to attempt it. Consider that the exposed (unclothed) areas of the body do most of the radiating. In addition to radiation loss from the body to the surrounding environment, there is also a radiation input to the body. Keep in mind, however, that although sunlight can help maintain body temperature, it cannot power the tissue-building process or fuel muscular work; these processes require chemical (food) energy.

EXERCISE 5:   How high can you climb on a pint of whiskey?

15. Debunking a Dynamo

Suppose somebody claims: "I have developed a heat engine that will operate by using the temperature difference between the top and bottom waters of a lake. It is a solar-powered device, since the sun's energy sustains this temperature difference. Using a lake that is 10^4 m^2 in area, and in which the temperature of the top and bottom waters are 25°C and 15°C, respectively, the engine will run for centuries. It will generate, on the average, a megawatt of electricity." Is the person telling the truth?

· · · · · · ·

When someone proposes a technical advance and you want to evaluate it (let alone invest in it!) it is sensible to check first whether it is compatible with the laws of nature. It is generally much easier to determine whether an hypothesized gadget violates physical principles (such as the law that information cannot be transmitted faster than the speed of light, or the law of conservation of energy) than to determine whether it will be unacceptable to society for reasons such as cost. This problem invites you to prove *unequivocally* that the engine promoter is a liar, not by showing that the proposed engine would be outrageously expensive or that it would require a backup system during a period when the lake was not stratified, but by identifying a fundamental law of nature that would be violated if the hypothesized engine existed.

Because the issue here is energy, you should look for a violation of the laws of thermodynamics. The first law states that energy is conserved, which assures us that the maximum sustained electric power output from the heat engine cannot exceed the rate at which solar energy strikes the lake surface. A megawatt of power produced over an area of 10^4 m^2 corresponds to a power density of 100 W/m^2. This is less than the average solar flux at Earth's surface, and thus the first law of thermodynamics is not violated.

But before you reach for your checkbook and invest in the device, make sure the second law is upheld. This law takes many forms. In this case, the clue to knowing which form to use is in the statement that the device is a heat engine (such engines are also called Carnot engines) operating between two given temperatures. The device takes in heat energy from the warmer portion of the lake, at the temperature $T_H = 25$°C, discharges some heat to the colder portion of the lake, at a temperature $T_C = 15$°C, and produces electricity at a rate equal to the difference between the rate of heat input and the rate of heat discharge.

The efficiency, ϵ, of such an engine, is defined as

$$\epsilon = \frac{\text{rate of electric energy output from engine}}{\text{rate of heat energy input to engine}}. \qquad (1)$$

Dividing both the numerator and denominator by the lake area, we can use Eq. 1 to express ϵ in terms of input and output power density.

The second law of thermodynamics states that the *maximum* efficiency, ϵ_{max}, of a Carnot engine is

$$\epsilon_{max} = \frac{T_H - T_C}{T_H}, \qquad (2)$$

when T_H and T_C are absolute temperatures measured in K. This yields a maximum efficiency of the engine of $(298 - 288)/298 = 0.034$.

At this point we know the maximum value of ϵ and the power density output as predicted by the inventor. Using these data and rearranging Eq. 1, we can calculate the minimum power density input required to run the engine and compare it with the actual power density input (which must be less than or equal to the solar flux on the lake if the engine is to work). In terms of power density and rearranged, Eq. 1 becomes

$$\text{minimum power density input} = \frac{\text{power density output}}{\epsilon_{max}} \qquad (3)$$

$$= \frac{100 \text{ W/m}^2}{0.034} = 2941 \text{ W/m}^2.$$

This minimum input is much higher than the daily averaged solar flux at Earth's surface, which rarely exceeds 500 W/m^2. The inventor is not telling the truth.

EXERCISE 1: Suppose that the inventor is more modest and claims to be able to produce only 0.02 megawatts of electricity (MWe). In this case, the second law doesn't forbid the device. What efficiency would the device be working at? [Use the value for $(\Omega/4)(1 - a)$, as given in Problem II.B.13, for average surface solar flux.] If the inventor claims the device will produce 0.02 megawatts for years and years, what technical problem should you then worry about and investigate before investing in the device?

16. Cooling Off Hot Plants

At what rate is water used to cool a 1000-megawatt coal-fired power plant?

· · · · · · ·

Most coal-fired electricity generating plants produce electricity by means of a steam turbine. The same is true, in fact, for oil, gas, nuclear, and many solar-powered electricity generating plants. The heat from the burning coal produces steam under pressure, which is directed at turbine blades, causing them to rotate. In this way, heat energy is converted to mechanical energy. The remaining step—conversion of mechanical energy to electricity—is accomplished by taking advantage of a remarkable property of electricity: When a conducting wire is moved across the flux lines of a magnetic field, electric current flows in the wire. To take advantage of this fact, the rotating turbine shaft is wrapped with conducting wires and surrounded by a strong magnetic field. The device, called a generator, produces an electric current.

The laws of thermodynamics tell us that the first step—conversion of heat to mechanical energy—cannot be, even in principle, 100% efficient, whereas the second conversion (from mechanical to electrical energy) can be. The maximum efficiency of the heat-to-mechanical energy conversion is given by the equation (see also Problem II.14)

$$\epsilon_{max} = \frac{T_H - T_C}{T_H},\qquad(1)$$

where the hotter temperature, T_H, is the temperature of the pressurized steam, and T_C is the temperature at which condensed, "spent" steam emerges from the turbine as liquid water. If $T_C = T_H$, the efficiency would be zero, which is another way of saying that if the emerging water is as hot as the steam entering the turbine, the turbine will not turn. The steam must be cooled and thereby depressurized at the back end of the turbine if there is to be a net pressure on the turbine blades, causing them to rotate.

The actual efficiency of such a power plant will be less than the ideal, ϵ_{max}. Inefficiencies in both conversion processes, resulting from mechanical friction, electrical resistance, loss of heat up the smoke stack, etc., result in energy loss.

For a typical coal-fired electric generating plant T_H is about 800 K, and T_C is about 300 K. The ideal efficiency, ϵ_{max}, is thus equal to 500/800 or 62.5%. The actual conversion efficiency for a modern coal-fired electricity generating plant is about 40%, which means that

$$\frac{\text{electric power output}}{\text{rate of heat input from coal burning}} = 0.40.\qquad(2)$$

The difference between the heat input and the electric output is waste heat. The rate at which this waste heat is produced, R, is

$$R = \text{(rate of heat input)} - \text{(rate of electrical energy output)}. \quad (3)$$

Using Eq. 2 we can rewrite this as

$$R = \frac{\text{(electric power output)}}{0.40} - \text{(electric power output)}. \quad (4)$$

For a 1000-megawatt (MWe) electricity generating plant,[21] this becomes

$$R = 1500 \text{ MW} \quad (5)$$
$$= 4.7 \times 10^{16} \text{ J/yr}.$$

Typically, about 15% of this waste heat is removed via the smoke-stack, in the form of hot effluent gases. The remaining 85%, or 4.0×10^{16} J/yr is the heat that must be discharged from the turbine by some cooling process in order to maintain T_C at 300 K (see Figure II-15). Three methods are available to accomplish this cooling:

(1) Once-Through Cooling In this process, cool water flowing past the power plant and through the turbine condensor is warmed by the waste heat and then discharged to the environment. Sea, river, or lake water is used for such a purpose. If the cooling water enters at, for example, an average temperature of 290 K, then it will typically be heated by about 10 K (to 300 K) as it passes through the turbine condensor. The rate at which such water must flow through the system can then be calculated. It takes 1 cal, or 4.18 J, to heat 1 g of water 1 K. The 4.0×10^{16} J ($= 9.6 \times 10^{15}$ cal) that must be discharged every year will heat 9.6×10^{15} g of water by 1 K, or 9.6×10^{14} g of water by the required 10 K. In other words, about 10^{15} g of water per year (32 m^3/sec) will flow through the turbine condensor and then be returned, warmed, to the local environment.

(2) Wet-Cooling Tower In this process, water is evaporated from the fins of a huge tower. Heat is expended to evaporate liquid water; this heat is removed from the site where the liquid water was, and therefore this site is cooled. Because the heat used to evaporate water is far more than one calorie per gram, a much smaller quantity of water is needed for this process than for once-through cooling. The water used is not returned to the local environment. Most of the evaporated water is likely to condense from vapor back to liquid and precipitate far downwind, and only a small fraction appears as fog in

---

21. The unit, MWe, is often used for the power output of an electricity generating plant. MWe stands for megawatts of electricity.

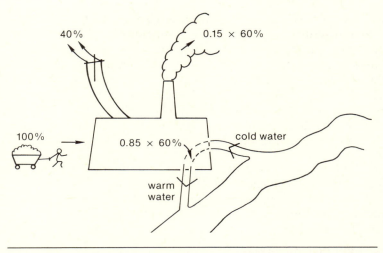

Figure II-15   The energy budget for a coal-fired electricity generating plant.

the vicinity of the cooling tower.[22] The rate at which water will be consumed in the wet-cooling-tower process roughly equals the rate of waste-heat removal, 9.6×10^{15} cal/yr divided by the number of calories per gram of water evaporated. The latter is called the heat of vaporization of water. The Appendix (VI.1), gives this constant as 587.7 cal/g if the liquid water is at a temperature of 290 K. Hence, 9.6×10^{15} cal/yr divided by 587.7 cal/g(H_2O), or 1.6×10^{13} g(H_2O)/yr must be consumed by this process for a 1000-MWe coal-fired electricity generating plant.

(3) Dry-Cooling Tower   In this process, the only water that is used flows in a closed cycle in the cooling tower. The scheme operates somewhat like the cooling system in an automobile, in which the water is cooled by air blown past the water in a radiator.

The environmental and economic trade-offs among these three processes are important. Once-through cooling results in thermal pollution—the returned water is 10 K warmer than the source water. Because the river or lake to which the water is returned will be warmer as a result of the thermal discharge, biological effects may occur. Also, the rate of evaporation from that water body will increase somewhat as a result of the higher temperature. Because the volume of water needed to flow past the power plant is large, large numbers of aquatic organisms in that water may be physically damaged during their passage through the cooling condensor. If an adequate year-round water supply is not present, a reservoir may have to be constructed, at potentially enormous ecological risks. The wet-cooling-tower approach is more costly and results in the local *consumption* of

---

22.   When the vapor later condenses back to liquid water, the heat used to evaporate the water reappears at the site of condensation.

water, which may be more harmful, environmentally, than the withdrawal and return of water to a local source. On the other hand, this cooling technique results in no discharge of heat to local waters (i. e., no thermal pollution). Moreover, the lower withdrawal requirements of the wet-tower method are less likely to necessitate reservoir construction. Dry cooling is the most expensive and the most environmentally benign of the three approaches, although it does slightly lower the overall efficiency of the power plant, thus necessitating the burning of more fuel to produce a given amount of electricity.[23]

EXERCISE 1: What percentage of the total annual runoff from the 48 coterminous states of the United States (the "lower 48") would have to be withdrawn for once-through cooling if all U.S. electricity demand in 1980 were to be met by coal-fired power plants using this cooling process?

EXERCISE 2: About 10 cm of precipitation falls each year in much of the dry southwestern United States, where new coal-fired power plants are being constructed. The rate at which water is consumed by a 1000-MWe coal-fired, wet-tower-cooled power plant corresponds to the precipitation rate over what land area?

EXERCISE 3: A nuclear power plant of the light-water reactor variety operates at about 32% efficiency. (The reactor is less efficient than a coal-fired plant because, for safety reasons, it uses a lower T_H.) Such a nuclear plant ejects practically none of its waste heat to the atmosphere, while in contrast a fossil-fuel fired plant ejects about 15% of its waste heat out the smokestack. For a given electrical power output, and for either wet-tower or once-through cooling, what is the ratio of the water requirement of the nuclear plant to that of the coal-fired plant?

EXERCISE 4: A once-through-cooled power plant would require far less cooling water if the water it did use were heated by much more than 10 K. What thermodynamic consideration tends to make this strategy unattractive?

EXERCISE 5: It takes energy to pump non-artesian groundwater up to the surface. Is it possible that an electrical power plant could require more water to cool it than could be pumped out of the ground by the plant's electrical output? To be specific, consider a wet-tower-cooled, coal-fired power plant and calculate the maximum water-table depth from which the cooling water requirement of the plant could be pumped if all of the electrical output were used only to pump groundwater for cooling. Assume that electric pumps can lift water with 90% efficiency.

---

23. These issues are discussed in more detail in Harte and El-Gasseir (1978).

17. When Waters Mix

If Mediterranean Sea water were allowed to flow into the Dead Sea, electric power could be derived in two ways. First, because the surface of the Dead Sea is about 400 m below the surface of the Mediterranean Sea, electrical energy could, in principle, be produced from falling water. Second, because on a weight-per-weight basis the Dead Sea is about 28% saline (only about 30% of which is common NaCl) and the Mediterranean Sea is about 3.8% saline, energy could, in principle, be derived from "entropic mixing." Suppose Mediterranean Sea water were to flow into the Dead Sea at a rate of 10^9 m³/yr and that the two energy conversion processes could be tapped at maximum thermodynamic efficiency. At what rate could electricity be derived from each?

· · · · · · ·

The notion of "entropic mixing" may be unfamiliar to you, but it will be explained shortly. Let's begin with the first method of deriving electric power because it is easier to understand. The rate at which electrical energy can be derived from falling water is deducible from Eq. 1 of Problem II.14. We need simply recall that the work needed to lift a mass, m, to a height, h, is equal to the energy that can be tapped when that mass falls the same vertical distance. The energy will be in the form of kinetic energy, which is convertible under ideal circumstances (no heat losses from, for example, friction) to electrical energy at 100% efficiency. In our problem the amount of energy that can be generated in one year by this gravitational source, E_{grav}, is given by

$$E_{grav} = mgh. \qquad (1)$$

Substituting numerical values,

$$E_{grav} = 10^9 \text{ m}^3(\text{H}_2\text{O}) \times \frac{10^3 \text{ kg}(\text{H}_2\text{O})}{\text{m}^3(\text{H}_2\text{O})} \times 9.8 \frac{\text{m}}{\text{sec}^2} \times 400 \text{ m}$$

$$= 3.9 \times 10^{15} \text{ kg m}^2/\text{sec}^2 = 3.9 \times 10^{15} \text{ J}.$$

Because this much electric energy is producible in one year, the rate of electricity production, p_{grav}, is

$$p_{grav} = \frac{3.9 \times 10^{15} \text{ J}}{3 \times 10^7 \text{ sec}} = 1.3 \times 10^8 \text{ W}. \qquad (2)$$

Calculation of the rate of production of energy from entropic mixing, p_{ent}, is a bit less straightforward. The idea is as follows. Entropy is a measure of how well mixed things are. Seawater is a mixture of salt and water (plus other chemicals that we can ignore here). It has a higher entropy than the water and salt would have if they were separated from each other. One form of the second law of thermodynamics states that it requires energy to reduce the entropy of a system. This is the reason why a desalination plant requires energy to operate: It converts high-entropy seawater into a lower entropy state consisting of pure water plus salt. By the same token, the mixing of salt and water (or water of two different salinities) can generate energy.

How large is the effect? Let's first take a little detour to calculate the minimum amount of energy needed to remove salt from a sample of salt water. One procedure for doing so consists of pushing the salt out with a semipermeable membrane through which water, but not salt, passes (see Figure II-16). The salt ions in the water behave as the molecules of a dilute gas; they obey the gas law

$$PV = n\mathrm{R}T, \tag{3}$$

where P equals the pressure of the ions (called the ionic pressure), V is the volume of the sample of seawater, n is the number of moles of salt ions present, R is a fundamental physical constant equal to 8.31 J per mole per K, and T is the temperature in K.

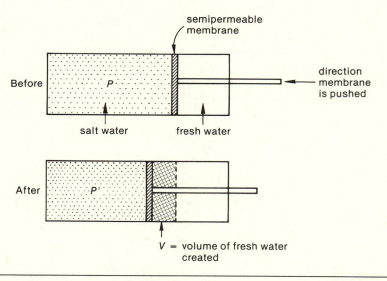

Figure II-16 Idealized representation of a desalination apparatus. As the piston with the semipermeable membrane is pushed to the left, salt and water are separated. If the volume of salt water is large compared to the volume of fresh water created, then the ion pressure, P, before will be nearly equal to the ion pressure, P', after. Hence, the amount of work needed to push the piston in will equal PV.

Consider 1 m^3 of typical seawater, which has a salt content of 35 parts per thousand or 3.5% (by weight) and a density of 1.025×10^3 kg/m^3. This sample will have a mass of 1,025 kg while the mass of salt ions in the cubic meter of seawater will be 1,025 x 0.035 or 36 kg. Suppose the only salt in the water is common table salt, NaCl. NaCl has a molecular weight of about 58.5, and there will be one mole of Na$^+$ and one mole of Cl$^-$ for every mole of NaCl. Thus[24]

$$n = 2 \times \frac{36 \times 10^3 \text{ g/m}^3}{58.5 \text{ g/mole(NaCl)}} = 1.23 \times 10^3 \text{ moles(Na}^+, \text{Cl}^-)/\text{m}^3. \quad (4)$$

Therefore, at the average surface temperature on Earth, 290 K,

$$P = \frac{\left(1.23 \times 10^3 \dfrac{\text{moles}}{\text{m}^3}\right) \times \left(8.31 \dfrac{\text{J}}{\text{(mole)(K)}}\right) \times (290 \text{ K})}{1 \text{ m}^3}$$

$$= 2.96 \times 10^6 \text{ J/m}^3. \quad (5)$$

Since a newton-meter is a joule, this can be rewritten as a more familiar expression for pressure in units of force per unit area:

$$P = 2.96 \times 10^6 \text{ N/m}^2. \quad (6)$$

Atmospheric pressure at Earth's surface is 1.013×10^5 N/m^2, so the pressure we have calculated is 29 times that of the atmosphere at sea level. In other words, the pressure of salt ions in seawater would be approximately 29 times the pressure of the atmosphere at sea level if all the salt were NaCl.

From Eq. 6 an approximate value for the minimum energy needed to desalinate seawater can be calculated. The amount of work required to push against a pressure, P, through a volume, V, is PV. Therefore, to clear the ions out of a cubic meter of seawater by pushing a semipermeable membrane through the water sample requires 2.96×10^6 J/m^3 $\times$ 1 m^3 or 2.96×10^6 J. This is equivalent to 2.9 J/g (seawater).

Why is this the *minimum* energy required to desalinate water? If a semipermeable membrane is pushed through an isolated cubic meter of seawater, then the salt concentration will gradually increase on the forward side of the membrane. Eqs. 5 and 6 will then be incorrect, because the concentration will continually increase forward of the membrane, exceeding the value of 3.5% salt concentration on which the calculation was based. The higher salt concentration will lead to a greater pressure, thus requiring more work to push it out of

---

24. Seawater actually contains other salts besides NaCl, so Eq. 4 is only approximate (see Exercise 1).

the way with the membrane. The value calculated here, 2.96×10^6 J/m^3, is an ideal thermodynamic limit. It is most realistic when the total volume of seawater (of which the cubic meter sample to be desalinated is a part) is very large. In that case, the salt ions pushed forward by the membrane are diluted by a large volume of water, and the increase in pressure is negligible.[25]

Now back to our problem. The difference in ionic pressure between the Mediterranean Sea water and the Dead Sea water can, in principle, be used to produce energy. A number of schemes for doing so are described in the technical literature [see, for example, Olsson et al. (1979) and references therein]. Details about the scheme are not important to us, though, because we only want to know the theoretical *maximum* rate at which energy can be derived. A thermodynamic process like desalination that uses energy requires the least energy when the process is carried out slowly in a reversible manner. Similarly, the amount of energy that can be derived from a process such as the mixing of solutions with different salinities is maximum when that process is carried out slowly and in a reversible manner. Because the mixing of solutions with different salinities is the reverse of the process of desalination, the *maximum* rate at which energy can be derived from the mixing of solutions equals the *minimum* energy needed to unmix those solutions. Hence, we can use the result above on the minimum energy required for desalination to derive the maximum energy that can be obtained from entropic mixing of waters with different salinities.

The difference in salt concentration between the Mediterranean and Dead Sea waters is 28% − 3.8%, or about 24%. This is 24/3.5 or 6.9 times the difference between the concentration of ions in seawater and fresh water. Using our calculated seawater ionic pressure of 29 atmospheres, the Mediterranean–Dead Sea ionic-pressure difference is 29 × 6.9, or 199 atmospheres.[26]

Next, we have to determine the rate at which energy can be derived from this pressure difference. To do this, we will calculate the height to which a column of water could be raised by the energy the pressure difference represents. One atmosphere of pressure corresponds to the pressure of a column of fresh water of height, h, where

$$\rho g h = 1.013 \times 10^5 \text{ N/m}. \tag{7}$$

---

25. For further discussion of this, see Harte and Socolow (1971). Van Ness (1969) gives a good general discussion of the conditions under which ideal thermodynamic limits apply.

26. The proportions of the various salt species in the Dead Sea are different from those in the Mediterranean Sea or in typical ocean waters. Moreover, as salt concentration gets large, ionic pressure is not simply linearly proportional to that concentration. For these reasons, the estimate of 199 atmospheres and the calculation in Eq. 9 are only approximate.

In this expression, ρ is the density of water (10^3 kg/m^3) and g is the acceleration of gravity (9.8 m/sec^2). Solving Eq. 7 for h, we obtain

$$h = \frac{1.013 \times 10^5}{9.8 \times 10^3} = 10.3 \text{ m.} \tag{8}$$

Therefore, a pressure difference of 199 atmospheres corresponds to a water-column height of 199 × 10.3 or 2,048 meters, which is 2,048/400 or 5.1 times as high as the elevation difference between the two seas. In other words, 5.1 times more power can, in principle, be derived from entropic mixing of these two waters than directly from falling water as calculated in Eq. 2. Hence

$$
\begin{aligned}
p_{ent} &= 5.1 \times p_{grav} \\
&= 5.1 \times 1.3 \times 10^8 \text{ W} \\
&= 6.6 \times 10^8 \text{ W.}
\end{aligned}
\tag{9}
$$

Actually, tapping the energy from entropic mixing is much more difficult than tapping energy directly from the elevation difference. Membranes exist that can accomplish the task of converting a salinity difference into a pressure difference, but they cannot function at the ideal thermodynamic limit we have calculated. By the same token, if water were to be channeled and tunneled from the Mediterranean to the Dead Sea, frictional losses along the way would reduce the power derived from that source as well. What we have calculated here is an ideal thermodynamic limit for each of the two processes. Technology today is far closer to tapping p_{grav} than p_{ent} at an efficiency close to the thermodynamic limit.

EXERCISE 1: Correct Eqs. 4 and 5, by including the total Na, Cl, Mg, and S concentrations of seawater (see Appendix (VIII)).

EXERCISE 2: Water bodies like the Dead Sea have a high salt concentration because they have no outlet. Water flows into them, in the form of runoff bearing salts from the surrounding land, and exits mainly by evaporation. This exiting water is extremely pure (distilled). Thus, salts are left behind and become quite concentrated over time. If the Jordan River, historically the main inflow to the Dead Sea, is to be partially diverted for irrigation water, leaving Mediterranean inflow as the Dead Sea's major water source, what factors would govern the maximum allowable rate of inflow of Mediterranean water? First consider maintenance of water volume. The area of the Dead Sea is about 1,050 km^2. (*a*) Using the fact that 1.5 m of water evaporates from the Dead Sea's surface each year, determine what fraction of the Jordan River inflow would have to be diverted if there is inflow of Mediterranean water at a rate of 10^9 m^3/yr and water volume remains constant. (*b*) Next consider salt buildup in the Dead Sea. The volume of water in the Dead Sea is 136 km^3. In how many years would inflow of Mediterranean seawater at a rate of 10^9 m^3/yr result in a doubling of the salt content of the Dead Sea?

18. Bouncing Sunbeams

A layer of clouds has a light reflection coefficient (albedo) of 0.5 and a transmission coefficient of 0.4. It lies above a patch of Earth's surface with an albedo of 0.1. The air between clouds and earth transmits solar radiation perfectly. What is the albedo of the combined cloud-surface system?

. . . .

When solar radiation strikes a layer of material in the atmosphere, it can undergo three possible processes. Some of the incident flux will be reflected back to outer space from the material; some will either be transmitted unaffected through the material or will scatter off the material without a change in wavelength; and some will be absorbed in the material and converted to infrared radiation. Denote the fractions of the incident flux that are reflected, transmitted, and absorbed by symbols R, T, amd A, respectively. The sum of $R + T + A$ is equal to one because reflection, transmission, and absorption are the only fates that can befall the incident light. R, the reflection coefficient of the layer, is often called the albedo; T is the transmission coefficient; and A is the absorption coefficient. All three are called optical coefficients.

Layers of material above or below a given layer also reflect, transmit, and absorb light. A pair of layers will bounce light back and forth between them like kids playing catch. However, some of the throws (to pursue the analogy) are not returned (i.e., not reflected) because the ball is either missed (i.e., transmitted) or caught and transformed into another condition (i.e., absorbed). To calculate the total reflection, transmission, or absorption by the two layers requires a mathematical summation of a series of sequential effects. The procedure used is the multiple scattering method illustrated below for the case where the top layer is a cloud layer and the bottom one is Earth's surface (see Figure II-17).

First let's estimate upper and lower bounds to the answer. The albedo of the combined system is the ratio of the rate at which light is reflected back to space to the rate at which light is incident on the combined system. Because the top layer has an albedo of 0.5 and the combined system cannot reflect less light than the top layer does, we expect that the albedo of the combined system will be greater than 0.5. A lower bound is thus 0.5. The rate at which light is reflected from Earth's surface is the albedo of that surface times the rate at which light hits it. Because the top layer both prevents some of the incident light from striking Earth's surface and also absorbs some of the light reflected from that surface (so that not all the reflected light

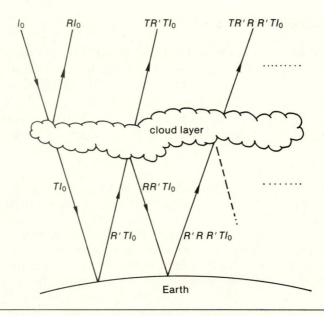

Figure II-17 The multiple scattering of the solar flux between Earth's surface and a cloud layer. I_0 is the incoming flux; R and R' are the reflection coefficients of the layer and of Earth's surface, respectively; T is the transmission coefficient of the layer.

reaches outer space), the albedo of the clouds and earth combined should be less than $0.5 + 0.1 = 0.6$. An upper bound is therefore 0.6. To obtain more precise values than these we must set up the problem properly.

Let R, T, and A be the optical coefficients for the top layer (clouds) and R', T', and A' be the optical coefficients for the bottom layer (Earth's surface). According to the problem statement, $R = 0.5$ and $T = 0.4$; thus $A = 1 - 0.5 - 0.4 = 0.1$. Because no light penetrates through Earth's surface, $T' = 0$. Since $R' = 0.1$, A' must equal 0.9.

Consider, now, an incident solar flux of intensity I_0. On its first pass through the cloud layer, a flux of strength RI_0 will be reflected directly back to outer space and a flux of strength TI_0 will be transmitted to Earth's surface. Of the flux TI_0 transmitted to Earth's surface on the first pass, a fraction, R', will be reflected and A' absorbed. Thus, a flux of strength $R'TI_0$ will hit the cloud layer from below. At this point we have to assume that the optical properties of the cloud are the same whether measured from the top or the bottom (see Exercise 4). Assuming this, a portion of this upward flux, of magnitude $TR'TI_0$, will be transmitted through the cloud to outer space and will contribute, along with that first reflected flux, RI_0, to the albedo of the combined system. At this point, the reflected flux is $RI_0 + TR'TI_0$.

As Figure II-17 shows, a fraction, R, of the upward flux, $R'TI_0$, is reflected back down to Earth's surface. Again, some of this is absorbed and some reflected back up. Continuing in this way, the total flux, F, sent back to outer space is[27]

$$F = RI_0 + TR'TI_0 + TR'RR'TI_0 + TR'RR'RR'TI_0 + \ldots . \quad (1)$$

This infinite sum can be written in the form

$$F = RI_0 + TR'[1 + (RR')^1 + (RR')^2 + \ldots]TI_0. \quad (2)$$

A commonly used notation for the term in brackets is

$$[1 + (RR')^1 + (RR')^2 + \ldots] = \sum_{i=0}^{\infty} (RR')^i. \quad (3)$$

In general, the following relation holds:

$$\sum_{i=0}^{\infty} (X)^i = \frac{1}{1 - X}, \quad (4)$$

if X is less than 1. Because optical coefficients like R and R' are fractions, they, and therefore their product, are less than 1. So, we can write

$$F = RI_0 + \sum_{i=0}^{\infty} (RR')^i \, T^2R'I_0$$

$$= RI_0 + \frac{T^2R'I_0}{1 - RR'} . \quad (5)$$

The albedo of the combined system is the ratio of the total flux reflected to outer space, F, to the incident flux, I_0. Thus, dividing the above by I_0, we obtain

$$\text{combined albedo} = R + \frac{T^2R'}{1 - RR'} . \quad (6)$$

Upon insertion of numerical values, this yields

$$\text{combined albedo} = 0.5 + \frac{(0.4)^2(0.1)}{1 - (0.5)(0.1)} = 0.52.$$

---

27. We follow here the approach of Rasool and Schneider (1971).

EXERCISE 1: For our rough guess, we argued that the combined albedo should be greater than or equal to R and less than or equal to $R + R'$. Show that, indeed, Eq. 6 implies a combined albedo that satisfies these inequalities for any values of the optical coefficients R, R', and T.

EXERCISE 2: Suppose a cloud layer with the same optical coefficients as above ($R = 0.5$, $T = 0.4$) lies above a region of Earth with a surface albedo of 0.8 instead of 0.1. What is the combined albedo of the cloud-surface system? Referring only to this albedo effect, consider how reflecting material in the atmosphere influences Earth's surface temperature. In particular, how does this influence differ between the polar regions and the temperate zone? What considerations other than albedo effects might be relevant to this difference?

EXERCISE 3: Derive general formulas for the fraction of the incident solar flux (*a*) absorbed at Earth's surface (f_S) and (*b*) absorbed in the cloud layer (f_C), and evaluate those quantities for the special case in which $R = 0.5$, $T = 0.4$, and $R' = 0.1$. Show that, in general, the total reflected flux plus the total absorbed flux equals the incident flux.

EXERCISE 4: Suppose two cloud layers (or a cloud layer and a layer of particles in the atmosphere) lie one above the other. One has optical coefficients of R, T, and A and the other of R', T', and A'. Ignore Earth's surface and considering the two layers as an isolated system (or, equivalently, let the absorption coefficient of Earth's surface equal 1), calculate the albedo of the two-layer system for two cases: (*a*) with the light encountering first the unprimed (R, T, A) system, with the primed (R', T', A') system behind it, and (*b*) the other way around. Does the albedo of the combined system depend on which layer the incident flux strikes first? Without doing any calculations, show how you could have reached your conclusion by considering a limiting case. To solve Problem II.18, we assumed that the optical coefficients of the cloud layer were the same, whether viewed from the top or the bottom. What does the result of this exercise tell you about the conditions under which that assumption would be invalid?

* EXERCISE 5: Pollution of the atmosphere with carbon soot particles of roughly a micron or smaller diameter is a major concern today.[28] In areas of the world once thought to be free of pollution, such as the Arctic, the atmospheric concentration of these fine particles of carbon soot is comparable to that in some urban areas of the United States. Arctic soot is believed to be produced by combustion

---

28. See, for example, Rosen et al. (1981).

of fossil fuel in the northern hemisphere. Soot in the atmosphere absorbs sunlight. If the soot lies above a region of Earth with high albedo (such as a region with low-lying clouds or the icy Arctic), it lowers the overall albedo of the region (see Exercise 2) by absorbing some sunlight before it gets a chance to reflect off the shiny surface below and by allowing less of the sunlight that penetrates to the surface to reach outer space again after reflection off the shiny material below. This tends to warm the Earth. On the other hand, if the high soot absorbs much of the incident sunlight, then less reaches the surface and the surface cools down. If the soot lies below the reflecting materials in the atmosphere, then the overall decrease in albedo should be less than if it lies above them.

To explore the interplay of these possibilities, consider a region of Earth with a surface reflection coefficient, R'', and an absorption coefficient, $A'' = 1 - R''$. Let the atmosphere have a reflection coefficient, R', equal to 0.26 and an absorption coefficient, A', of 0.23. (Both these values are good averages for Earth's atmosphere.) Finally, let a layer of particulates with a reflection coefficient, R, of zero and an absorption coefficient, A, reside in the atmosphere. Consider the following cases:

	High particles			Low particles		
	$A = 0$	$A = 0.1$	$A = 0.5$	$A = 0$	$A = 0.1$	$A = 0.5$
$R'' = 0.16$						
$R'' = 0.5$						

The case of the high particles can be treated by letting the unprimed layer lie above the primed; for the low-particle case, these are reversed. The case $R'' = 0.16$ corresponds to the globally averaged surface reflectivity, while an R'' of 0.5 is typical for the north polar region. For all 12 cases, calculate (a) the value of the total albedo, a, of the region, and (b) the fraction of the region's incident sunlight that is absorbed at the surface.

From your results, what can you say about the effects of surface albedo and particulate altitude on the changes in total albedo and fraction of sunlight absorbed at Earth's surface? Remember that when $A = 0$, no particles are present. Would high or low particles be more likely to warm the Earth?

Later on, after reading Problem III.6, you will have access to a climate model that will enable you to analyze quantitatively the effect of the interplay of albedo and surface absorption on surface temperature.

EXERCISE 6: Our treatment in Problem II.18 of multiple scattering between a cloud layer and the surface of Earth involved one major oversimplification, alluded to in the opening paragraph. As you know from watching the reflected glare of light off a roadway, the albedo of a surface depends on the angle between the incident rays and the surface. Perpendicular rays are more likely to be absorbed than are rays at a grazing angle. The optical coefficients given in Problem II.18 apply to an average incident solar flux angle. Notice, too, that we assumed light scattered off the cloud layer would be transmitted unaffected through the layer. In fact, however, the scattering of incident light by a cloud layer alters the average angle of the light striking Earth's surface and thus alters the effective albedo.

What qualitative effect do you think our oversimplification had on the answer? Include in your deliberations on this question the fact that the absorption probability in the cloud layer is proportional to the path-length of light through the layer. This is relevant because light reflected more obliquely off Earth's surface will have a greater path-length within the cloud on its subsequent pass upwards. On the other hand, because the cloud surface is relatively rough, the reflection coefficient of the cloud layer is not as dependent on incident angle as that of Earth's surface is.

C.
Chemical Reactions and Equilibria

When chemicals are added to pure water, reactions can occur that cause changes in the types and concentrations of chemical species in solution. Usually within seconds or minutes, except in special cases such as when slow dissolution of a solid occurs, a steady state is reached in which the concentration of each chemical species is constant. This is called chemical equilibrium.

Suppose chemical species **A** and **B** are added to water and undergo the reaction

$$\mathbf{A} + \mathbf{B} \rightleftharpoons \mathbf{C} + \mathbf{D}, \tag{1}$$

where the double arrow indicates that the reaction can go either way. At chemical equilibrium, the forward reaction, $\mathbf{A} + \mathbf{B} \rightarrow \mathbf{C} + \mathbf{D}$, occurs at the same rate as the reverse reaction, and a simple relation must hold among the concentrations of the species. Using the standard notation [**X**] to indicate the concentration of chemical species **X**, in units of moles per liter, this relation reads

$$\frac{[\mathbf{C}][\mathbf{D}]}{[\mathbf{A}][\mathbf{B}]} = 10^{-pK}, \tag{2}$$

where pK is a constant characterizing the specific reaction. The term 10^{-pK} is often called a "dissociation constant" for the reaction. Values of 10^{-pK} for a number of reactions of environmental interest are given in the Appendix (X.2). A major problem in aqueous chemistry is the determination of equilibrium concentrations. In the above example, a single equation contains four unknowns. How is the problem to be solved?

Most such problems fall into one of two categories. Either (*a*) the system is completely closed to the atmosphere and to surrounding solids, in which case certain conservation laws can be used; or (*b*) the system is open to the surroundings and a combination of conservation laws and across-the-phase-boundary equilibrium relations can be used. These two methods are illustrated by the following examples.

CASE A: In a container of pure water, a certain amount of raw nitric acid is dissolved. The water + acid system is capped and thus isolated from the surroundings. (The container walls are assumed to be inert.) What will be the steady-state concentration of H^+, OH^-, HNO_3, and NO_3^-?

CASE B: Into a container of water open to the atmosphere at sea level, a hunk of $CaCO_3$ (limestone) is dropped. What will be the steady-state concentration of H^+, OH^-, H_2CO_3, HCO_3^-, and Ca^{+2}?

One important relation that will be applicable in both these cases and, in fact, is used in all aqueous equilibrium problems, stems from consideration of the reaction: $H_2O \rightleftharpoons H^+ + OH^-$. The forward reaction is called a dissociation reaction because water dissociates into its ionic products. The reverse reaction is called an association reaction. Eq. 2 can be applied here, of course, and because the *pK* for the dissociation of water[29] is 14,

$$[H^+][OH^-] = 10^{-14}. \tag{3}$$

The convention adopted in writing Eq. 3 is that $[H_2O] = 1$, and the *pK* value of 14 is assigned with that convention in mind.

Armed with this background, you are now ready to follow the procedures for analyzing cases A and B.

Case A

When nitric acid is added to water, the following reaction occurs:

$$HNO_3 \rightleftharpoons H^+ + NO_3^-. \tag{4}$$

The *pK* for this reaction is -1 and thus, in equilibrium

$$[H^+][NO_3^-] = 10 \, [HNO_3]. \tag{5}$$

In addition, an important conservation law can be applied. The amount of NO_3 in the system is constant because the system is

---

29.   The *pK* of pure-water dissociation is 14 only at 25°C. Chemical equilibrium constants, such as the *pK*'s, Henry's constants, and solubility products, are all temperature dependent. A discussion of this and of most other aspects of aquatic chemistry as well is found in the excellent text by Stumm and Morgan (1981).

closed. The NO_3 may be in ionic form (NO_3^-) or in undissociated form (HNO_3), but the sum total of the two forms must equal the amount of NO_3 added initially. Therefore,

$$[NO_3^-] + [HNO_3] = A, \tag{6}$$

where A can be calculated from the amount of acid initially added. For example, if 126 g of pure HNO_3 were added to 1 liter of water, the number of moles of NO_3 added would be 126 g divided by the molecular weight of HNO_3, which is 63. A would then be 2 moles per liter. If the same amount of acid were added to 4 liters of water, the value of A would be 0.5 moles per liter. If 2 cc (2×10^{-3} liters) of a 3 mole/liter solution of HNO_3 were added to 5 liters of water, then A would be $0.002 \times 3/5 = 0.0012$ moles/liter. In these examples, a small correction was ignored: after the acid is added, the volume of the liquid will be slightly greater than before. In the last example, the correct answer would actually be obtained by dividing 0.006 moles by $(5 + 2 \times 10^{-3})$ liters, a small correction, indeed.

So now we have four unknowns (H^+, OH^-, NO_3^-, and HNO_3) and three equations (3, 5, and 6). The final equation (necessary if we are to maintain our faith in chemical determinism) is obtained from the principle of charge conservation, which tells us that the total concentration of positive charge must equal the total concentration of negative charge. The reason the conservation law takes this form is that we started with pure, neutral water and neutral nitric acid and so we must end up with an electrically neutral system as well.

The charge conservation equation is:

$$[H^+] = [OH^-] + [NO_3^-]. \tag{7}$$

The left- and right-hand sides of this equation are the total number of moles per liter of positive and negative charge, respectively. The four equations (3, 5, 6, and 7) are sufficient to determine the four unknown concentrations. The best procedure to be used for solving the four equations will depend on the actual concentrations, because various kinds of approximations usually have to be made. We'll discuss this as we solve the problems later in this Section.

Case B

This example involves two sources of soluble chemicals. CO_2 from the atmosphere will come into solution in the water and $CaCO_3$ will dissolve from the limestone and produce dissociated Ca^{+2} and CO_3^{-2}. In the water, the CO_2 and the CO_3^{-2} will take part in the reactions

$$H_2CO_3 \rightleftharpoons H^+ + HCO_3^-$$

(remember, H_2CO_3 includes dissolved CO_2), and

$$HCO_3^- \rightleftharpoons H^+ + CO_3^{-2}.$$

At 25°C, the pK of the first reaction is 6.35 (often denoted pK_1) and that of the second is 10.33 (pK_2), so that

$$[H^+][HCO_3^-] = 10^{-6.35} [H_2CO_3] \tag{8}$$

and

$$[H^+][CO_3^{-2}] = 10^{-10.33} [HCO_3^-]. \tag{9}$$

The value of $[H_2CO_3]$ is obtained from the application of Henry's law. According to this law, when water is in contact with the atmosphere for a sufficient time to allow gas exchange between air and water to equilibrate (reach a point of no net flow), the ratio of the aqueous concentration of a gas to the atmospheric concentration of that gas is a constant (called Henry's constant) that depends only on temperature but not on the actual concentrations. Like many similar-looking laws, this one breaks down if the concentrations get too high; but for the problem at hand this presents no concern. Henry's law can be written for our case:

$$[H_2CO_3] = p(CO_2)K_H, \tag{10}$$

where $p(CO_2)$ is the atmospheric concentration of CO_2 at sea level (340 ppm) and K_H is Henry's constant for CO_2. In units such that aqueous concentrations are expressed in moles per liter and atmospheric concentrations are expressed in atmospheres [10^6 ppm(v)], K_H for CO_2 has the value $10^{-1.47}$ (at 25°C), and so

$$[H_2CO_3] = 340 \times 10^{-6} \times 10^{-1.47} = 10^{-4.94}. \tag{11}$$

We now have four equations (3, 8, 9, and 11) for the six unknowns. Another equation is the charge conservation relation:

$$[H^+] + 2[Ca^{+2}] = [OH^-] + [HCO_3^-] + 2[CO_3^{-2}]. \tag{12}$$

Note the factor of 2 in front of the calcium and carbonate terms in this equation. Do you see why it is there?

The final equation needed before we can solve for our unknowns involves the steady-state relation between the solid limestone and the dissolved Ca^{+2} and CO_3^{-2}. Every solid has a solubility product that tells you how much of the solid can dissolve before precipitation back to the solid phase occurs. For limestone, the solubility product is $10^{-8.42}$, expressed in units of (moles/liter)^2 (see Appendix, X.3). Thus, at saturation,

$$[Ca^{+2}][CO_3^{-2}] = 10^{-8.42}. \qquad (13)$$

To solve our problem, we must assume either (*a*) that the hunk of limestone was large enough to permit saturation to take place (before the hunk dissolved completely) or (*b*) that the hunk did dissolve completely and the amount of $CaCO_3$ in it was known precisely at the outset. Eq. 13 results if the first of these two assumptions is true. An equation analogous to Eq. 6 results if the second assumption is true, namely

$$[Ca^{+2}] = \text{specified constant.} \qquad (14)$$

In either case, with six equations and six unknowns, the system can be solved. We'll use these tools in Problems II.20 and 21, and III.1, 2, and 4.

19. Altering the Atmosphere by Burning Fossil Fuels

In 1980, how much O_2 was removed from the atmosphere due to the combustion of fossil fuel on Earth, and how much CO_2 and H_2O were produced in the combustion process?

· · · · · · · ·

The primary elemental constituents of the three major types of fossil fuel (natural gas, petroleum, and coal) are carbon and hydrogen. When fossil fuel is burned, oxygen from the atmosphere combines with the carbon to make CO_2 and with the hydrogen to make H_2O. In addition, coal contains some water (typically 10–15% by weight), which is released to the atmosphere upon combustion of the coal. All three types of fossil fuel contain various other substances such as ash, sulfur, and trace metals in even lower concentrations.

The Appendix (VII.2–4) provides detailed information about the average chemical composition of the three fossil fuels, the heat value of a unit amount of each fuel, and the total amount of heat energy derived worldwide from each in 1980. To use this information to compute the rates of CO_2 and H_2O emission and of oxygen consumption, we need to balance the chemical combustion reactions for each fuel. Consider petroleum, first, with the approximate chemical composition $CH_{1.5}$. The combustion reaction for this fuel is

$$CH_{1.5} + xO_2 \rightarrow yCO_2 + zH_2O, \tag{1}$$

where x, y, and z are called stoichiometric constants. Their values, to be determined below, provide the answer to our problem: For each mole of petroleum consumed, x moles of oxygen are consumed, and y and z moles, respectively, of CO_2 and H_2O are produced. The number of moles of petroleum burned in 1980, in turn, can be determined from the data in the Appendix (VII.2–4).

Let's first determine the stoichiometric constants. Equating moles of C, H, and O on each side of the reaction reveals constraints on x, y, and z. One mole of carbon in $CH_{1.5}$, for example, produces y moles of C in the form of CO_2, and so

$$y = 1. \tag{2}$$

Similarly, one and a half moles of H in $CH_{1.5}$ produces z moles of H_2 or $2z$ moles of H, and so

$$2z = 1.5 \tag{3}$$

or

$$z = 0.75. \tag{4}$$

Finally, $2x$ moles of O produces $2y$ moles of O in CO_2 and z moles of O in H_2O, or

$$2x = 2y + z. \tag{5}$$

Combining Eqs. 2, 4, and 5, we get

$$x = 1.375. \tag{6}$$

The number of moles of $CH_{1.5}$ burned worldwide in 1980 now has to be determined. From the Appendix (VII.2–4), we learn that the energy content of the combusted petroleum was 1.35×10^{20} J; petroleum has a heat content of 4.3×10^{10} J/tonne. By weight, petroleum is 98% $CH_{1.5}$; thus, the amount of $CH_{1.5}$ consumed in 1980 was

$$0.98 \times \frac{1.35 \times 10^{20} \text{ J}}{4.3 \times 10^{10} \text{ J/tonne}} = 3.08 \times 10^9 \text{ tonnes}(CH_{1.5})$$

$$= 3.08 \times 10^{15} \text{ g}(CH_{1.5}). \tag{7}$$

Since one mole of $CH_{1.5}$ has a mass of $12 + 1.5$ or 13.5 g, the number of moles of $CH_{1.5}$ consumed in 1980 was

$$\frac{3.08 \times 10^{15}}{13.5} = 2.28 \times 10^{14} \text{ moles}(CH_{1.5}). \tag{8}$$

Hence $n(O_2)$, the number of moles of consumed O_2, was x times this value, or

$$n(O_2) = 1.375 \times 2.28 \times 10^{14}$$

$$= 3.14 \times 10^{14} \text{ moles}(O_2). \tag{9}$$

The values of $n(CO_2)$ and $n(H_2O)$ can be obtained by multiplying 2.28×10^{14} moles$(CH_{1.5})$ by y and z, respectively. Hence,

$$n(CO_2) = 2.28 \times 10^{14} \text{ moles}(CO_2) \tag{10}$$

and

$$n(H_2O) = 1.71 \times 10^{14} \text{ moles}(H_2O). \tag{11}$$

A similar procedure is used for calculating $n(O_2)$, $n(CO_2)$, and $n(H_2O)$ from coal and natural gas combustion. However, now the more diverse chemical nature of the fuel must be taken into account. Referring to the Appendix (VII), we find that the composition of natural gas, expressed as mole fractions, is CH_4 (75%), C_2H_6 (6%), C_3H_8 (4%), C_4H_{10} (2%), and C_5H_{12} (1%). The remaining 12% in noncombustible. Using these mole fractions of each hydrocarbon in natural gas, you can show that a mole of that fuel contains [0.75(1) + 0.06(2) + 0.04(3) + 0.02(4) + 0.01(5)] moles of C and [0.75(4) + 0.06(6) + 0.04(8) + 0.02(10) + 0.01(12)] moles of H. Thus, the effective formula of the combustible portion of the fuel is $C_{1.12}H_4$.

The stoichiometric constants can now be determined as before, resulting in

$$C_{1.12}H_4 + 2.12O_2 = 1.12CO_2 + 2H_2O. \tag{12}$$

The number of moles of natural gas burned in 1980 will equal the total heat value derived (6.0×10^{19} J; see Appendix, VII.2) divided by the number of joules per cubic meter of gas (3.9×10^7 J/m^3; see Appendix, VII.4), and then multiplied by the number of moles per cubic meter (44.6 moles/m^3 for any gas, since one mole of any gas occupies 22.4 liters at standard temperature and pressure). This works out to be 6.9×10^{13} moles of natural gas. Multiplying by 0.88 (the combustible fraction) and by the approximate stoichiometric constants gives the number of moles of O_2 consumed and of CO_2 and H_2O produced.

For coal, the only subtlety is that 13% of coal is water, which is liberated to the atmosphere upon combustion. This must be included in the calculation.

The final result for all three fuels is conveniently summarized in the following table (see Exercise 1).

	$n(CO_2)$*	$n(H_2O)$*	$n(O_2)$*
Petroleum	2.28	1.71	3.14
Natural gas	0.68	1.21	1.28
Coal	1.80	0.94	2.16
Total	4.76	3.86	6.58

*($\times$ 10^{14} moles)

EXERCISE 1: Derive the results for natural gas and coal shown in the table above.

EXERCISE 2: If all the CO_2 released to the atmosphere in 1980 from fossil-fuel burning remained there, by what percentage would that year's CO_2 emission increase the atmospheric concentration?

EXERCISE 3: By what percentage did the O_2 consumed in 1980 by fossil-fuel burning deplete the atmosphere's stock of O_2?

EXERCISE 4:   There is great concern that the CO_2 we add to the atmosphere from fossil-fuel burning will result in a so-called "greenhouse" effect because that gas traps outgoing infrared radiation and thereby warms Earth's surface (see Problem III.8). We will see later (also in Problem III.8) that H_2O is a more effective absorber of infrared radiation than is CO_2. Given that emissions of H_2O were comparable to those of CO_2, why don't we worry about the effect of our H_2O emissions on the radiation balance in the atmosphere? (Hint: answering this requires only stock-flow–residence time considerations.)

20. The pH of Pristine Precipitation

What would the pH of rain be in the absence of anthropogenic sources of sulfuric and nitric acids?

. . . .

In the absence of sulfur and nitrogen oxide emission from fuel burning, Earth's atmosphere would still contain N_2, O_2, water vapor, dust, and trace gases of natural origin such as CO_2, SO_2, and NH_3. A raindrop falling through Earth's pristine atmosphere will absorb these gases to some extent. If, dissolved in water, these gases form acids or bases, the pH of the rain will be affected. Geological history suggests that pristine precipitation very likely was acidic. The weathering processes that slowly convert many common hard minerals to clays result from chemical reactions between precipitation runoff and mineral substrate. These are usually acid-induced reactions.

Since CO_2, dissolved in water, forms carbonic acid (H_2CO_3), let's look quantitatively at that one first. The method outlined in the introduction to this section is applicable. Using the 1983 value of 340 ppm(v) for the concentration of CO_2 in Earth's atmosphere, and the values of Henry's constant and the dissociation constants from the Appendix (X.2, 3),

$$[H_2CO_3] = 10^{-4.94}, \tag{1}$$

$$[H^+] [HCO_3^-] = 10^{-6.35} [H_2CO_3], \tag{2}$$

and

$$[H^+][CO_3^{-2}] = 10^{-10.33} [HCO_3^-]. \tag{3}$$

In addition to these equilibrium equations, we have the equilibrium relation for water:

$$[H^+][OH^-] = 10^{-14}. \tag{4}$$

Eq. 1 is valid provided the raindrop is at sea level, where CO_2 has a partial pressure of 340×10^{-6} atm. At higher altitude, where the partial pressure of CO_2 is lower, $[H_2CO_3]$ will be proportionately lower as well.

The charge conservation equation reads

$$[H^+] = [HCO_3^-] + 2[CO_3^{-2}] + [OH^-], \tag{5}$$

and now there are five equations for five unknowns. Usually the best way to solve the equations in chemical equilibrium problems is to start with the charge conservation equation, rewritten with the term you want to solve for on the left-hand side and all the others on the right. We want to know $[H^+]$, so we've written Eq. 5 in this form. The next step is to replace the terms on the right-hand side with functions of $[H^+]$, alone. This is accomplished by substitution, using the equilibrium relations. In particular for our problem, if Eq. 1 is substituted into Eq. 2 we obtain

$$[H^+][HCO_3^-] = 10^{-11.29} \tag{6}$$

or

$$[HCO_3^-] = \frac{10^{-11.29}}{[H^+]}, \tag{7}$$

giving us $[HCO_3^-]$ written as a function of $[H^+]$. Using Eq. 7, Eq. 3 becomes

$$[H^+][CO_3^{-2}] = \frac{10^{-21.62}}{[H^+]} \tag{8}$$

or

$$[CO_3^{-2}] = \frac{10^{-21.62}}{[H^+]^2}. \tag{9}$$

Now $[CO_3^-]$ is a function of $[H^+]$, alone. Using Eq. 4 we determine that

$$[OH^-] = \frac{10^{-14}}{[H^+]}. \tag{10}$$

Substituting Eqs. 7, 9, and 10 into Eq. 5, we arrive at an equation involving only $[H^+]$:

$$[H^+] = \frac{10^{-11.29}}{[H^+]} + \frac{10^{-21.32}}{[H^+]^2} + \frac{10^{-14}}{[H^+]}. \tag{11}$$

This cubic equation can be solved, now, for $[H^+]$ [for the procedure, see Birkhoff and MacLane (1953)]. Instead of solving for the exact solution, however, we will employ a simple method for finding an approximate solution. Since this method is often applicable in chemical equilibrium problems, it is worth learning.

To start, let's suppose that the first term on the righthand side of Eq. 11 is much larger than the other two terms—so much greater that, essentially,

$$[H^+]^2 = 10^{-11.29} \tag{12}$$

or

$$[H^+] = 10^{-5.65}. \tag{13}$$

If this provisional value for $[H^+]$ is substituted back into the right-hand side of Eq. 11, the three terms, reading from left to right, equal $10^{-5.65}$, $10^{-10.33}$, and $10^{-8.35}$. The first of these is, in fact, considerably larger than the other two, which tells us that our hypothesis was consistent with the correct solution. Summarizing the method (see also Figure II-18): By hypothesizing that two terms were much smaller than the third we were led to an equation (Eq. 12) whose solution indicates that the two terms were, indeed, smaller. Remembering that $pH = -\log_{10} [H^+]$, we determine the pH of rain in equilibrium with atmospheric CO_2 to be $-\log_{10} 10^{-5.65} = 5.65$.

Another acid-forming gas is SO_2. The reactions $H_2SO_3 \rightleftharpoons H^+ + HSO_3^-$ and $HSO_3^- \rightleftharpoons H^+ + SO_3^{-2}$ govern the dissociation of dissolved SO_2. Using the two values of the dissociation constants, Henry's constant from the Appendix (X.3), and an approximate background concentration of SO_2 in Earth's atmosphere of 0.2 ppb(v), the equilibrium relations are:

$$[H_2SO_3] = (0.20 \times 10^{-9} \text{ atm}) \times (10^{0.096} \text{ moles/liter-atm}) \tag{14}$$
$$= 10^{-9.60} \text{ moles/ liter,}$$

$$[H^+][HSO_3^-] = 10^{-1.77} [H_2SO_3], \tag{15}$$

and

$$[H^+][SO_3^{-2}] = 10^{-7.21} [HSO_3^-]. \tag{16}$$

Given: $X = \dfrac{a}{X} + f(X)$ 1.

Assume: $X \approx \dfrac{a}{X}$ or $X \approx \sqrt{a}$ 2.

Substitute 2 into 1:
$$\sqrt{a} \approx \frac{a}{\sqrt{a}} + f(\sqrt{a}) \qquad 3.$$

3 and, therefore, 2 are valid if
$$f(\sqrt{a}) << \sqrt{a}$$

Figure II-18 A procedure for finding an approximate solution to the chemical equilibrium problem.

These relations are correct regardless of how much CO_2 is dissolved in the rain; similarly Eqs. 1–3 are correct regardless of how much SO_2 is dissolved in the rain. The relation that links CO_2 and SO_2 is the charge balance equation, which for the combined $CO_2 + SO_2$ systems reads

$$[H^+] = [HCO_3^-] + 2[CO_3^{-2}] + [HSO_3^-] + 2[SO_3^{-2}] + [OH^-]. \quad (17)$$

Using Eqs. 7, 9, 10, and 14–16, this becomes

$$[H^+] = \frac{10^{-11.29}}{[H^+]} + \frac{10^{-21.32}}{[H^+]^2} + \frac{10^{-11.37}}{[H^+]} + \frac{10^{-18.26}}{[H^+]^2} + \frac{10^{-14}}{[H^+]}. \quad (18)$$

Supposing that the first and third terms on the righthand side are bigger than the others, we calculate that

$$[H^+]^2 = 10^{-11.29} + 10^{-11.37} \quad (19)$$

or

$$[H^+] = 10^{-5.51}. \quad (20)$$

The pH of the rain is thus 5.51. Substitution of Eq. 20 into Eq. 18 verifies that the second, fourth, and fifth terms were indeed small compared to the first and third.

The pH we have calculated should be corrected for two effects ignored up to this point. First, ammonia gas in the atmosphere forms a base when dissolved in water. Exercise 3 deals with this correction. Second, alkaline dust is present in the atmosphere over certain continental regions of Earth, such as the Great Basin of the United States (the land between the Sierra Nevada and the Rocky Mountains). Precipitation falling in the Rocky Mountains is sometimes found to be acidic (pH $\cong$ 4.5) owing to human activities, but occasionally it has a pH greater than 6.0. The rain or snow storms with high pH usually contain $CaCO_3$ (limestone) or $CaMg(CO_3)_2$ (dolomite). Exercise 4 will give you practice in applying equilibrium chemistry to limestone dissolution in raindrops.

Because the "background" pH of precipitation should be in the mid fives, only precipitation with pH around five or below is now referred to as "acid precipitation." For a further discussion of this issue, with a treatment of background SO_4^{-2}, see Charlson and Rodhe (1982).

EXERCISE 1: Before the industrial revolution, the concentration of CO_2 in Earth's atmosphere was about 275 ppm(v). Considering the effect of dissolved CO_2 only, calculate the effect that the increase in CO_2 has had on the pH of precipitation.

EXERCISE 2: Here's a chance to practice solving some cubic equations by trial and error. (*a*) Consider the equation: $x^3 - 5x = 125$. Which of the following yields the best approximate solution to this equation: the x^3 term is small and $-5x \approx 125$; the $5x$ term is small and $x^3 \approx 125$; 125 is small and $x^3 \approx 5x$? (*b*) What is an approximate positive solution to $x^3 + 30x^2 - 10x - 40 = 0$?

* *EXERCISE 3:* In raindrops, the reaction $NH_3 + H_2O \rightleftharpoons NH_4^+ + OH^-$ occurs, which tends to elevate pH. Henry's constant and the NH_3 dissociation constant are given in Sections X.2, 3 of the Appendix. The actual influence of background ammonia in the atmosphere on precipitation pH is difficult to estimate because the atmospheric NH_3 concentration, $p(NH_3)$, is poorly known. The Appendix (Section V.2) only provides an upper limit on $p(NH_3)$. One clue about the value of $p(NH_3)$ is provided by the observation that in remote areas, the measured NH_4^+ concentration in rainwater is generally less than about 5 micromoles per liter. Show that in the presence of atmospheric CO_2 at 340 ppm and SO_2 at 0.2 ppb a rainwater value of $[NH_4^+] < 5 \times 10^{-6}$ implies an upper limit on $p(NH_3)$. What is that upper limit? What will the pH of pristine precipitation be with $p(CO_2)$ equaling 340 ppm(v), $p(SO_2)$ equaling 0.2 ppb(v), and $p(NH_3)$ equaling 0.01 ppb(v)?

* *EXERCISE 4:* A sample of rain water is observed to have a pH of 7.4 (*a*) If only atmospheric CO_2 at 340 ppm(v) and limestone dust are present in the atmosphere to alter the pH from a neutral value of 7, and if each raindrop has a volume of 0.02 cm^3, what mass of Ca is present in each raindrop? (*b*) Suppose there was so much $CaCO_3$ in the raindrops that the rainwater was saturated. What would the pH of the raindrops be?

* *EXERCISE 5:* Show that the residence time for atmospheric SO_2 with respect to its dissolution and subsequent removal through precipitation (as dissolved H_2SO_3, HSO_3^-, and SO_3^{-2}) is independent of the atmospheric concentration of SO_2 and depends only on the pH of the precipitation and the precipitation rate. At a pH of 5.51 (Eq. 18) and a precipitation rate of 1 m/yr, what is the residence time? In a pristine atmosphere, what fraction of the SO_2 emitted to the atmosphere from natural sources would be removed via this rain-out process?

21. Natural Acidity from Biological Processes

The nitrification process is a source of acidity. In typical freshwater lakes, the nitrification rate averages about 10^{-4} moles (N)/liter/yr. Compare the influx of H^+ to a lake receiving inflow water with a pH of 4 (presumed due to acid rain) with the influx of H^+ from the internal nitrification process in the lake. Assume the residence time of water in the lake is half a year.

. . . .

The rate of inflow of H^+ to the lake in stream water, $F_{H,S}$, equals the rate of inflow of stream water, F_W, times the concentration of H^+ in that water, $[H^+]$. The rate of inflow of water equals the lake volume, V, divided by the residence time, T. Hence, we can state

$$F_{H,S} = F_W [H^+] \tag{1}$$
$$= \frac{V[H^+]}{T}.$$

With V in liters, $[H^+]$ in moles per liter, and T in years, the rate of inflow of H^+ will be in units of moles of H^+ per year. Substituting numerical values for $[H^+]$ and T, we obtain

$$F_{H,S} = \frac{10^{-4}\ V}{0.5} = 2 \times 10^{-4}\ V. \tag{2}$$

To calculate the influx of H^+ from internal nitrification, $F_{H,N}$, we turn to Section X.1 of the Appendix to find the nitrification reaction:

$$NH_4^+ + 2O_2 \rightarrow NO_3^- + 2H^+ + H_2O. \tag{3}$$

The structure of this reaction tells us that for every mole of NH_4^+ that is nitrified to NO_3^-, two moles of H^+ are produced. Hence, if 10^{-4} moles(N)/liter are nitrified each year, then 2×10^{-4} moles(H^+)/liter will also be produced each year. Multiplying this by the number of liters of water in the lake, V, we get an inflow of H^+ from nitrification of

$$F_{H,N} = 2 \times 10^{-4}\ V, \tag{4}$$

in units of moles/yr. Thus, the two inflow rates are equal.

EXERCISE 1: (*a*) Using the data in Sections XII.1 and XIII.2 of the Appendix, estimate the total annual amount of H^+ produced on the continents in the nitrification process. (*b*) What pH of precipitation on the continents would yield an equal flux of H^+?

EXERCISE 2: Given the result above, and given that hydraulic residence times for lakes are often larger than half a year, why should we be worried about acid damage to lakes only from precipitation? After all, it appears that a natural biological process contributes as much acidity as does acid rain.

D.
Non-Steady-State Box Models

The problems in Section A were solved by equating compartment in-flows to compartment outflows. This was appropriate because the problems involved steady-state situations. A more difficult class of problems involves stocks that change over time, a situation arising when inflows are not in balance with outflows. When the inflow, F_{in}, is not equal to the outflow, F_{out}, then the rate of change of the stock is given by

$$\text{rate of change of } M = F_{in} - F_{out}. \tag{1}$$

If $F_{in} > F_{out}$, M increases with time; while if $F_{in} < F_{out}$, M decreases. In either case, there is no longer a well-defined residence time.

Eq. 1 is the starting point for many non-steady-state box models. If the flows are known functions of the stock, then the equation can be solved for $M(t)$. The problems in this section illustrate how this is done for the case in which the stocks vary smoothly, so that a differential equation

$$\frac{dM}{dt} = F_{in} - F_{out} \tag{2}$$

can be written.

In a multi-box model, a *set* of equations like Eq. 1 results. If the flow in or out of one box depends upon the stocks in that box and also the stocks in other boxes, then a set of interconnected, or cou-pled, equations results. The treatment of the interconnected boxes in a non-steady-state model is more complicated and is deferred to Chapter III (Problems III.5 and 14).

22. Exhausting Fossil Fuel Resources (II)

If the present worldwide rate of consumption of petroleum increases by 2% every year, how long will it take to use up Earth's petroleum resource?

· · · · · · ·

In this problem the non-steady-state quantity in the box is Earth's petroleum resource. The statement that consumption increases by a fixed percentage every year implies that the resource is being removed at an exponentially increasing rate. Of course, it is not being added to at all. Thus, the instantaneous consumption rate, or outflow, $F(t)$, is an exponential of the form

$$F(t) = F(0) e^{rt}, \tag{1}$$

where $t = 0$ refers to the present. Using the 1980 value for the present rate of petroleum consumption (see Appendix, VII.2), $F(0) = 1.35 \times 10^{20}$ J/yr. The rate constant, r, in units of $(yr)^{-1}$ can be determined by the condition that $F(1) = 1.02 F(0)$, or

$$e^{r(1)} = 1.02. \tag{2}$$

Because $\log_e (1.02) \approx 0.0198$,

$$r \approx 0.0198. \tag{3}$$

If $M(t)$ is the amount of petroleum remaining at time t [with $M(0) = 1.0 \times 10^{22}$ joules as given in the Appendix, VII.2] then the rate of change of $M(t)$ is given by

$$\frac{dM}{dt} = -F(0) e^{rt}. \tag{4}$$

This equation can be integrated to yield an equivalent equation. If the integration variable is called t' and if the integral extends from $t' = 0$ to $t' = t$, this equivalent equation takes the form

$$\int_{M(0)}^{M(t)} dM = -\int_0^t dt' F(0) e^{rt'}. \tag{5}$$

The integrals are easily carried out to yield

$$M(t) - M(0) = \frac{-F(0)}{r} e^{rt} + \frac{F(0)}{r}, \tag{6}$$

or,

$$M(t) = M(0) + \frac{F(0)}{r} (1 - e^{rt}). \qquad (7)$$

The petroleum resource will be entirely consumed when M vanishes. $M(t)$ will vanish at time T, where T is given by

$$M(0) + \frac{F(0)}{r} (1 - e^{rT}) = 0, \qquad (8)$$

or

$$e^{rT} = 1 + \frac{rM(0)}{F(0)}. \qquad (9)$$

Taking the natural log of both sides, we obtain

$$rT = \log_e\left[1 + \frac{rM(0)}{F(0)}\right]. \qquad (10)$$

Substituting the numerical values for r, $F(0)$, and $M(0)$ gives the final answer:

$$T = 45.6 \text{ yr.}$$

EXERCISE 1:   If coal and natural gas consumption are increasing at an annual rate of 4% per year and 1% per year, respectively, how long will Earth's resources of these fuels last?

EXERCISE 2:   Eq. 2 implies that the instantaneous rate of consumption at the end of a year is 2% larger than at the beginning of that year. Suppose we interpret the statement that the "rate of consumption of petroleum increases by 2% every year" to mean that the total consumption in year n is 2% greater than the total consumption in year $n - 1$. Using the fact that consumption in year n equals $\int_{n-1}^{n} F(t)\, dt$, show that Eq. 2 is still correct.

EXERCISE 3:   Problem II.22 can also be tackled using discrete time rather than continuous time. Let years be labeled with a symbol, n, that takes on values $n = 0$ (1980), $n = 1$ (1981), etc. If the amount of petroleum consumed in year n is $F(n)$, then $F(1) = 1.02\, F(0)$, $F(2) = (1.02)^2\, F(0)$, etc. By summing a finite geometric series, determine how big n must be so that $F(0) + F(1) + \ldots + F(n)$ roughly equals $M(0)$.

23. Pollution Buildup in a Lake

A lake has a volume of 10^6 m^3 and a surface area of 6×10^4 m^2. Water flows into the lake at an average rate of 0.005 m^3/sec. The amount of water that evaporates yearly from the lake is equivalent in volume to the lake's top meter of water. Initially, the lakewater is pristine, but at a certain time a soluble, noncodistilling pollutant is discharged into the lake at a steady rate of 40 tonnes/yr. Derive a formula for the concentration of pollutant in the lake as a function of time since the pollutant discharge began.

· · · · · · · ·

A noncodistilling pollutant is a substance that does not evaporate away with evaporating water. Therefore, evaporation of lakewater is not an exit pathway for the pollutant. However, if the lakewater flows out of the lake in an outlet stream or via underground seepage, that water outflow will remove pollutant. So to begin, let's calculate the stream and seepage outflow rate for the lakewater. The lakewater is in a steady state, with stock, M_w, equal to 10^6 tonnes(H_2O) (since 1 m^3 of water has a mass of 1 tonne). The water inflow rate, F_w, is 0.005 tonnes(H_2O)/sec = 1.6×10^5 tonnes(H_2O)/yr and therefore the total water outflow rate must also equal 1.6×10^5 tonnes(H_2O)/yr. The evaporation outflow rate is 1 m/yr times the area of the lake, or 0.6×10^5 tonnes(H_2O)/yr, and hence the stream and seepage outflow rate is 10^5 tonnes(H_2O)/yr.

At any time, t, let $M_p(t)$ equal the mass of pollutant in the lake in units of tonnes. The concentration of pollutant at time t is $M_p(t)/M_w$ in units of tonnes(pollutant)/tonne(water). If the stream-plus-seepage outflow rate is multiplied by the concentration of pollutant, and if the pollutant is well mixed in the lakewater the rate of pollutant outflow, $F_{p,out}(t)$, is obtained:

$$F_{p,out} = 10^5 \frac{M_p}{M_w}$$

$$= 0.1 \, M_p$$

(1)

in units of tonnes(pollutant)/yr.

The rate at which pollutant flows into the lake, $F_{p,in}$, equals 40 tonnes(pollutant)/yr. Equating the rate of change of M_p to the net difference between inflow and outflow,

$$\frac{dM_p}{dt} = F_{p,in} - F_{p,out}$$

$$= 40 - 0.1\,M_p, \tag{2}$$

with M_p in units of tonnes(pollutant) and time in units of years. This equation is of the form,

$$\frac{dX}{dt} = a + bX \tag{3}$$

and has a general solution[30]

$$X(t) = \frac{-a}{b} + ce^{bt}. \tag{4}$$

The constant, c, must be determined from a specified condition on $X(t)$. Thus if $X(0)$ is a known amount, the relation

$$X(0) = \frac{-a}{b} + c, \tag{5}$$

which follows from Eq. 4, determines the unknown constant, c. In our case, $a = 40$, $b = -0.1$, and, if $t = 0$ is the time the pollutant discharge began, then $X(0) = 0$. Therefore, $c = a/b$. In units of tonnes, the mass of pollutant in the lake is given by

$$M_p(t) = \frac{40}{0.1} - \frac{40}{0.1}\,e^{-0.1t}. \tag{6}$$

 EXERCISE 1:   What amount and what concentration of pollutant will exist in the lake as t approaches infinity?

 EXERCISE 2:   Since the water in the lake is in steady state, its residence time, T_w, can be computed. What is the value of that residence time? As t approaches infinity, the pollutant approaches a steady state; therefore a residence time, T_p, is derivable. What is the

---

30. You can derive the general solution by rewriting Eq. 3 in the form $dX/(a + bX) = dt$ and integrating both sides. You can also verify that Eq. 4 is correct by direct substitution into Eq. 3. An excellent introduction to differential equations, with considerable emphasis on practical applications, is the text by Boyce and DiPrima (1973).

value of that residence time? Explain why T_p is greater than T_w, and interpret the difference $T_w^{-1} - T_p^{-1}$, using your answer from Exercise 3, Problem II.6.

EXERCISE 3: Draw a plot of Eq. 6 to get a visual sense of how pollution builds up in a lake. Include values of $X(t)$ at 2-year intervals for a 20-year period starting at $t = 0$.

*** EXERCISE 4:** This exercise will give you some practice deriving and solving a differential equation. Consider a slab of material of thickness, x, and area, A. One face of the slab, the hotter face, is kept at a fixed temperature, T_1. The opposite face is fixed at temperature, T_2. Heat will flow across the slab, from the hotter face to the cooler one. The formula for the rate of heat conduction across the slab is

$$\frac{dQ}{dt} = \frac{kA(T_1 - T_2)}{x},$$

where Q is heat, t is time, and k is a constant called the coefficient of heat conductivity. Imagine that the slab of material is a layer of ice on a lake. A is the area and x is now the thickness of the ice. The air temperature right above the ice is a constant $-30°C$ and the water temperature just below the ice is a constant $0°C$. Thus $T_2 = -30°C$ and $T_1 = 0°C$. To form new ice from liquid water at $0°C$, 80 calories must be removed from every gram of the liquid water. The value of k for ice is 2.2 J/m°C sec. If the ice is 1 cm thick at $t=0$, how thick will it be two months later? (Hint: you first have to write a differential equation for dx/dt.) The answer you derive is roughly the amount of ice that would form on many lakes in the northern hemisphere in the aftermath of a major nuclear war. This is discussed in more detail in an article on "nuclear winter" by Ehrlich et al. (1983); see also Exercise 4 of Problem III.6.

Chapter III
Beyond the Back of the Envelope

Problems worthy of attack, prove their worth by hitting back.

—Piet Hein

The problems in this chapter are more open-ended than those in Chapters I and II. Solving them requires the creation and application of mathematical models. If you haven't had much experience with it, you may think systems modeling is a well-defined, somewhat rigid procedure akin to building an erector-set likeness of a bridge. Actually, the process is a dynamic one, involving flexible and creative thinking that results in a flexible product amenable to further remolding and refining as the modeling process continues. A good problem-solving model affords an array of elegant revelations about a complex system or situation. With a good model, you can view a situation from many perspectives. By tinkering with the model, you can journey mentally to extreme conditions and thereby learn more about how the system works. (For example, you can ask what our weather would be like if urbanization covered Earth's surface—see Problem III.9.)

You will find that after you've created a model to answer a question, it will start to ask *you* questions—ones you wouldn't have thought of before. Models also have a way of letting you know when you've made them clumsy and useless—when you must return to the drawing board. This is a recognition I experienced many times while preparing this book's problems and solutions. Indeed, the models presented here can certainly be improved or refined further by the ambitious reader.

The most difficult aspect of creating models lies in discovering how much detail to include. Often a lot of detail can be ignored provided the right details are included. So the first task is really to discern the "essence" of the problem. This will involve some trial and error. To provide insight into this process, I have described some of the exploratory thinking that preceded development of the models used below. In addition, several of the homework exercises in this chapter will give you a chance to practice creating models on your own.

A.
Biogeochemistry

A variety of chemical elements and compounds flow from one location to another on Earth. These locations include the oceans, the atmospheres, soil, plants and animals, and fossil fuels. Such biogeochemical flows both sustain life and have the potential to destroy it. In some situations, inflow and outflow are nearly in balance (more precisely, the stock is large compared to the difference between annual inflow and annual outflow) so that the time scale for change in stocks is very long. In others, particularly where human activity has disrupted natural flows and stocks, relatively short time scales are involved.

The problems in this section deal with both natural and altered flows. Problems III.1, 2, 4, and 5 are concerned with changes in flows or stocks of substances resulting from major human intervention in biogeochemical cycles, while Problem III.3 focuses on a relatively undisturbed situation. The methods used for solving Problems III.1, 2, 3, and 4 were developed in Chapter II, but here the problems require the use of an assemblage of tools. For example, chemical equilibrium theory applied in a multi-box configuration is used in Problems III.2 and 3. Problem III.5 introduces a more advanced mathematical method that will likely be new to many of you. The method involves the use of matrices to solve a set of interconnected differential equations. That problem is intended for the ambitious among you.

119

1. Acid Rain

The best current estimate is that in the northeastern United States, roughly one fourth of the SO_2 emitted to the atmosphere is oxidized to sulfate and subsequently deposited as sulfuric acid in rain and snow within the region. In the northeastern United States what precipitation pH should result?

. . . .

First let's figure out how much sulfur, in moles, is deposited as dissolved sulfate in any of the three forms (H_2SO_4, HSO_4^-, SO_4^{-2}) in the northeastern United States in one year. Then we can estimate the volume of precipitation falling in the region in one year, from which we can calculate the concentration of dissolved sulfate. Finally, we will apply chemical equilibrium theory to calculate the pH of that solution.

In recent years, about 10^7 tonnes(S) are emitted in the region annually. One fourth, or 2.5×10^6 tonnes(S), is oxidized to sulfate and subsequently deposited with rain or snow in the region. Converting to moles, the amount of dissolved sulfate [2.5×10^6 tonnes(S)] $\times$ [10^6 g(S)/tonnes(S)] divided by 32 g(S)/mole(S) = 7.8×10^{10} moles(S).

Approximately 1.5 m of precipitation falls annually in the northeastern United States. Since the area of that region, as defined loosely in Problem II.5, is about 10^{12} m^2, a volume of 1.5×10^{12} m^3 or 1.5×10^{15} liters of precipitation falls each year. Hence the concentration of dissolved sulfur, in the three forms of sulfate, is 7.8×10^{10} moles(S)/1.5×10^{15} liters or about 5×10^{-5} moles(S)/liter.

To proceed, we use the dissociation constants for the reactions $H_2SO_4 \rightleftharpoons H^+ + HSO_4^-$ and $HSO_4^- \rightleftharpoons H^+ + SO_4^{-2}$ given in the Appendix (X.2) to write the equilibrium relations:

$$[H^+][HSO_4^-] = 10^3 [H_2SO_4] \qquad (1)$$

and

$$[H^+][SO_4^{-2}] = 10^{-1.9} [HSO_4^-]. \qquad (2)$$

In addition, we know that

$$[H^+][OH^-] = 10^{-14}. \qquad (3)$$

We can also write an approximate[1] charge-conservation relation in the form

$$[H^+] = [HSO_4^-] + 2[SO_4^{-2}] + [OH^-]. \tag{4}$$

Finally, we have the relation expressing the total concentration of the three forms of sulfur:

$$[H_2SO_4] + [HSO_4^-] + [SO_4^{-2}] = 5 \times 10^{-5} \tag{5}$$
$$= 10^{-4.3}.$$

We now have five unknowns and five equations. Using the same sort of strategy executed in Problem II.20, we will express all the terms in the charge conservation relation (Eq. 4) ultimately as functions of $[H^+]$. To do this, we will first use the three equilibrium relations (Eqs. 1–3) to express all the terms in Eq. 4 as functions of $[H_2SO_4]$ and $[H^+]$. We will than substitute Eq. 1 and an equation derived from Eqs. 1 and 2 into Eq. 5 to produce a second equation relating $[H^+]$ and $[H_2SO_4]$. At that point we will have two equations and two unknowns, $[H_2SO_4]$ and $[H^+]$, from which an equation for $[H^+]$ alone can be derived.

Proceeding as outlined above, $[SO_4]$ can be expressed in terms of $[H_2SO_4]$ by substituting Eq. 2 into Eq. 1 so that

$$[SO_4^{-2}] = \frac{10^{1.1}\,[H_2SO_4]}{[H^+]^2}. \tag{6}$$

Substituting Eqs. 1, 3, and 6 into Eq. 4 we get

$$[H^+] = \frac{10^3\,[H_2SO_4]}{[H^+]} + \frac{10^{1.4}\,[H_2SO_4]}{[H^+]^2} + \frac{10^{-14}}{[H^+]}. \tag{7}$$

Now we must derive a second relation between $[H^+]$ and $[H_2SO_4]$. This can be achieved by substituting Eqs. 1 and 6 into Eq. 5, yielding

$$[H_2SO_4]\left[1 + \frac{10^3}{[H^+]} + \frac{10^{1.1}}{[H^+]^2}\right] = 10^{-4.3} \tag{8}$$

or

$$[H_2SO_4] = 10^{-4.3}\left[1 + \frac{10^3}{[H^+]} + \frac{10^{1.1}}{[H^+]^2}\right]^{-1}. \tag{9}$$

---

1. HCO_3^- and CO_3^{-2} are ignored here—see Exercise 1.

If Eq. 9 is substituted into Eq. 7, a nonlinear equation for $[H^+]$ results. The equation could, in principle, be solved exactly, but, as before, we will look for an approximate analytical solution.

Assume that the 5×10^{-5} M of sulfur are entirely dissociated, so that H_2SO_4 is completely dissociated to $HSO_4^- + H^+$ and HSO_4^- is completely dissociated to $SO_4^{-2} + H^+$. Then for every mole of all three types of sulfate added, there are two moles of H^+ in solution, or

$$[H^+] = 2 \times 10^{-4.3} = 10^{-4}. \tag{10}$$

Is this consistent with Eqs. 7 and 9? If Eq. 10 is substituted into Eq. 9, we calculate that

$$[H_2SO_4] = 10^{-4.3}[1 + 10^7 + 10^{9.1}]^{-1} \tag{11}$$
$$= 10^{-13.4}.$$

If Eqs. 10 and 11 are substituted into Eq. 7,

$$[H^+] = \frac{10^3 \, 10^{-13.4}}{10^{-4}} + \frac{10^{1.4} \, 10^{-13.4}}{10^{-8}} + \frac{10^{-14}}{10^{-4}} \tag{12}$$
$$\approx 10^{-4}.$$

So, indeed, Eq. 10 is correct and the pH of the precipitation is

$$pH = -\log_{10}[H^+] = 4. \tag{13}$$

This value is actually a good round-number estimate of the average pH of precipitation in regions of the world (such as northern Europe and the northeastern United States) where acid precipitation is a serious problem. You may wonder what happens to the rest of the SO_2. Some of the SO_2 forms sulfate that is deposited in rain or snow in Canada, on the Atlantic Ocean, or in the southeastern United States. Much of the rest falls to earth in the form of dry particulate matter in a process called "dry deposition." This dry sulfate can form sulfuric acid when rain water eventually dissolves it. In an elegant pair of articles, Oppenheimer (1983a, 1983b) has applied box-model concepts to the analysis of regional acid precipitation formation and dispersal. His analysis shows that a reduction in SO_2 emissions in the eastern and midwestern United States will result in a proportional reduction in H^+ deposition in rain and snow in that area.

Now that we have worked out the answer the long and rigorous way, it is illuminating to go back and look at a shortcut. In paragraph three of our discussion, above, we estimated the molar concentration of S in precipitation to be 5×10^{-5} moles(S)/liter. When H_2SO_4 dissociates to $HSO_4^- + H^+$ and HSO_4^- dissociates to $SO_4^{-2} + H^+$, two moles of H^+ are produced for every mole of H_2SO_4. Therefore, if all the H_2SO_4 and HSO_4^- that are formed dissociate completely, the 5×10^{-5} moles (S)/liter will produce $2 \times 5 \times 10^{-5} = 10^{-4}$ moles(H^+)/liter. We can therefore obtain the correct answer with very little effort provided the H_2SO_4 completely dissociates to $HSO_4^- + H^+$ and the HSO_4^- completely dissociates to $SO_4^{-2} + H^+$. This complete dissociation will occur if the pH that finally results is higher than the pK's of the dissociation reactions.[2] In our case, the pK's are -3 and 1.9, which are, indeed, considerably lower than the pH of 4 that results. The shortcut is therefore justified. To apply it in a new situation, first assume complete dissociation of the acid or acids present and thereby quickly calculate a pH. If that pH exceeds the pK's then the pH you derived is the correct one.

EXERCISE 1: It is probably intuitively clear to you that you will not alter very much the pH of a solution that has a pH of 4.0 by adding something that if dissolved by itself in pure water would produce a pH of 5.6. Verify this by showing that atmospheric CO_2 can be ignored in the calculation above.

EXERCISE 2: The result in Eq. 13 does not include the presence of nitric acid in precipitation in the northeastern United States. The precipitation actually contains about 1 mole of H^+ from nitric acid for every 2 moles of H^+ from sulfuric acid. What should the pH of the precipitation be with this correction?

EXERCISE 3: If one fourth of all the anthropogenic SO_2 emitted globally were to be converted to sulfuric acid and mixed uniformly in global precipitation, what would the pH of that precipitation be?

EXERCISE 4: Referring to the shortcut procedure described in the last paragraph of the problem solution, what would you do by way of a shortcut if the pH came out to be -1, a value lower than one of the pK's ($pK = 1.9$) but not the other?

---

2. This can be derived from Eqs. 1–4 of this problem.

2. Mobilization of Trace Metals

Acid precipitation can dissolve, or mobilize, certain trace metals, such as aluminum, from soils and lake sediments. This has occurred in many of the lakes of the Adirondack mountains of New York state, where dissolved Al^{+3} is now found in elevated concentrations that are toxic to some of the lakes' organisms. The same aluminum, locked up in a solid substrate, would be relatively harmless. What concentration of acid-mobilized trace metals should we expect to find in acidified lakes that lie in watersheds receiving precipitation with a pH of 4.0?

·······

As a rough estimate of the equilibrium aluminum concentration in the lake, assume that the concentration will be one third of the hydrogen ion concentration in the precipitation (since, by charge conservation, three moles of H^+ are required to mobilize one mole of Al^{+3}). Then, $[Al] = (1/3) \times 10^{-4.0}$ or $10^{-4.48}$ moles of Al^{+3} per liter. This should be an underestimate of the correct value because it ignores evaporation, which will concentrate dissolved substances such as H^+ in the drainage and Al^{+3} in the lake. On the other hand, it could be an overestimate because it doesn't take into account the possible pH-dependence of the mobilization process.

To develop a model, we must be more specific, so let's concoct an illustrative watershed with an area of 1 km^2, draining into a lake with a volume of 10^4 m^3. During a typical year, 2 m of precipitation, with a pH of 4.0, fall in the watershed. One third of this precipitation evaporates rapidly from the watershed soil surface, and two thirds penetrate the soil and enter the lake. The rate of outflow from the lake equals one half the rate of inflow, with the balance made up by evaporation from the lake surface. We will ignore aluminum deposited in the watershed by precipitation.

If the watershed is already acidified, we can assume that the bicarbonate buffering capacity of the watershed has been exhausted. In fact, we will take as the only[3] acid-neutralizing mechanism the mobilization (or dissolution) of Al^{+3} from watershed soils and lake sediments by the H^+ in the precipitation. Charge conservation tells us that three H^+ ions will mobilize one Al^{+3} ion; in the process the three H^+ ions will be removed from the water. In that sense, mobilization

---

3. This is, of course, a simplification. Other metals, such as iron, manganese, copper, and cadmium can be acid-mobilized as well. Moreover, not all mobilized aluminum is in the ionic form, Al^{+3}.

of metals neutralizes acidity. This mobilization process is pH-dependent; the greater the H^+ concentration, the more aluminum is mobilized. In particular, when aluminum is mobilized from a commonly found solid form, $Al(OH)_3$, the approximate relation,[4]

$$\frac{[Al^{+3}]}{[H^+]^3} = 10^{8.5} \tag{1}$$

holds. Here $[Al^{+3}]$ is the eventual equilibrium concentration of dissolved Al^{3+}, $[H^+]$ is the eventual equilibrium concentration of H^+, and the assumed reaction is $3H^+ + Al(OH)_3 \rightarrow Al^{+3}$ and $3H_2O$.

Deriving the correct answer now will be expedited by thinking sequentially about the flow of water and dissolved ions through the system (see Figure III-1). First, let's calculate what the pH of the water entering the lake would be if there were no Al^{+3} in the soil to partially neutralize it. Then we can worry about what happens when that inflow water enters the lake.

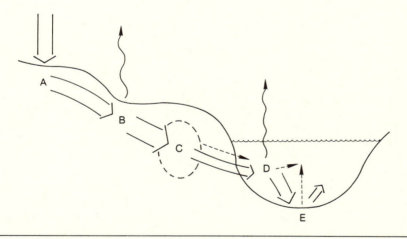

Figure III-1 The stages of metal mobilization in an acidified watershed. At A, acid precipitation enters the watershed. At B, evaporation of water (undulating arrow) concentrates the H^+ in the water, and so the solid arrow leaving B is thicker than the one entering. At C, aluminum is mobilized and flows (dashed arrow) into the lake. The process reduces the acidity of the water, as indicated by the thinner solid arrow leaving C. At D, evaporation from the lake increases both the metal and acid concentration of the lake water. At E, further mobilization takes place from the lake sediment.

---

4. This equation is an equilibrium relationship for the ionic concentrations in the aqueous environment of a large supply of $Al(OH)_3$. The equilibrium constant, $10^{8.5}$, and those describing acid-induced mobilization of other metals, are given in the Appendix (X.5). Stumm and Morgan (1981, pp. 238–49 and 538–48) provide more detailed information about the pH dependence of the dissolution of trace metals in aqueous solutions.

The concentration of H^+ in the precipitation is $10^{-4.0}$ moles/liter. But one third of the precipitation evaporates, leaving an enhanced concentration of $10^{-4.0}$ divided by 2/3, or $10^{-3.82}$ (because the same amount of H^+ will be found in two thirds the volume of water). Therefore the pH of the water would be 3.82.

But the pH of the water entering the lake will not be this small, because Al^{+3} will replace H^+ until the equilibrium relation, Eq. 1, holds. If Eq. 1 is supplemented with a charge conservation relation[5]

$$[H^+] + 3[Al^{+3}] = \text{constant}, \tag{2}$$

then the equilibrium value of $[H^+]$ and $[Al^{+3}]$ can be calculated. The constant in Eq. 2 is the value of $[H^+]$ when $[Al^{+3}] = 0$, or $10^{-3.82}$ moles/liter. When Eqs. 1 and 2 are combined, a cubic equation for $[H^+]$ or for $[Al^{+3}]$ is obtained, with an approximate solution (see Exercise 1)

$$[H^+] = 10^{-4.32} \text{ moles/liter} \tag{3}$$

and

$$[Al^{+3}] = 10^{-4.46} \text{ moles/liter}.$$

These are the values of $[H^+]$ and $[Al^{+3}]$ in the water entering the lake. In the lake, evaporation concentrates the H^+ and the Al^{+3}. The water outflow rate equals one half the water inflow rate, with evaporation making up the difference; thus, evaporation, by itself, will lead to lakewater concentrations in the steady state that are double those of the inflow.[6] Hence the hydrogen ion concentration in the lakewater would be (were it not for further mobilization of aluminum from lake sediments) $2 \times 10^{-4.32}$ or $10^{-4.02}$ moles/liter. The aluminum concentration would be $2 \times 10^{-4.46}$ or $10^{-4.16}$. Because at these concentrations the equilibrium relation, Eq. 1, would not be satisfied, further mobilization would occur. Before mobilization from lake sediment occurs, the numerical value of the total charge concentration, $[H^+] + 3[Al^{+3}]$, is $10^{-4.02} + 3 \times 10^{-4.16}$, or $10^{-3.52}$, and this must again remain constant by charge conservation. So for the lakewater concentrations of H^+ and Al^{+3} we again have two relationships:

$$[H^+] + 3[Al^{+3}] = 10^{-3.52} \tag{4}$$

---

5.   There will be other charged chemical species present, such as the anion of the acid, but they do not participate appreciably in the mobilization process.
6.   See Problem II.6 if this is not immediately obvious to you.

and Eq. 1. Solving these, we obtain

$$[H^+] = 10^{-4.20}$$

and

$$[Al^{+3}] = 10^{-4.10} \tag{5}$$

for the equilibrium *lake* concentrations.

Actual values of $[H^+]$ and $[Al^{+3}]$ in acidified lakes in the Adirondacks and Scandinavia are lower by a factor of 2 or more. A major reason is that aluminum mobilization is not really the only acid-neutralizing process occurring in the watershed.

Note that we didn't need the information concerning the watershed area, the lake volume, and the rate of precipitation to the watershed! An empirical study of acid-precipitation-induced aluminum mobilization in the northeastern United States is described in Cronan and Schofield (1979).

EXERCISE 1:   Show, by direct substitution, that Eq. 3 is a solution to Eqs. 1 and 2.

EXERCISE 2:   What are the units of the term, $10^{8.5}$, in Eq. 1?

EXERCISE 3:   How many parts per million (by weight) is $10^{-4.10}$ moles(Al^{+3}) per liter?

EXERCISE 4:   At what rate, in units of g(Al^{+3})/yr, will Al^{+3} be exported out of the watershed in the lake outflow?

EXERCISE 5:   Assuming the crustal abundance of aluminum (see Section VIII of the Appendix, VIII) is applicable to soil and sediment in the watershed, and assuming that aluminum is mobilized primarily from the top 20 cm of soil and sediment, how long can the mobilization process in the illustrative watershed continue before the aluminum in the upper soil layer and lake sediment is depleted? Assume a crustal density of 2 g/cm^3 in the top 20 cm.

* *EXERCISE 6:*   The fact that our answer was independent of the three factors—watershed area, lake volume, and precipitation rate—is a result of an unstated assumption we made concerning rates of reactions and rates of water flow. What was that assumption and what would be the qualitative effect on our answer of relaxing it? In particular, how would the answer depend on the three factors?

3. Tracing the Carbon Cycle

The ratio of dissolved C^{14} to C^{12} in the deep Pacific Ocean is about 0.77 times that in the atmosphere, and about 0.81 times that in the surface waters. Using C^{14} as a tracer, (a) estimate the residence time of dissolved inorganic carbon in the Pacific Ocean and (b) estimate the rate of exchange of CO_2 between the atmosphere and this ocean.

· · · · · · · ·

This problem shows how to use a radioactive isotope as a sort of clock to help us learn how long carbon lingers in the sea and to estimate its rate of flow along a major portion of its cycle. To begin, let's understand why the ratio of C^{14} to C^{12} is lower in seawater than in the atmosphere. The difference must involve the balancing of sources (inflows) and sinks (outflows) of C^{14} and C^{12} in the ocean and in the atmosphere.

Consider the properties of C^{14} and radioactive decay. Most carbon atoms have an atomic mass of 12; this form of carbon is denoted C^{12}. A relatively rare carbon isotope, denoted C^{14}, has an atomic mass of 14. Its formation in the atmosphere is initiated by cosmic radiation. CO_2, of both the $C^{12}O_2$ and the $C^{14}O_2$ variety, enters the oceans by gaseous diffusion and also diffuses back to the atmosphere from the ocean.

Both in seawater and the atmosphere, C^{14} decays at an exponential rate with a half-life $(T_{1/2})$ of 5,700 years. This means that in any large sample of C^{14} atoms, half will have decayed and half will remain after 5,700 years. Exponential radioactive decay, quite generally, is described by a mathematical formula

$$N(t) = N(0)\, e^{-\lambda t}, \tag{1}$$

where $N(t)$ is the number of radioactive atoms present at time t. $N(0)$ is the initial number present at time $t = 0$ [which is arbitrary; $t = 0$ could stand for 11:33 am (GMT), June 21, 1983, or any other time you choose to begin counting]. λ is the decay rate constant. The rate of decay of N is given by

$$\frac{dN}{dt} = -\lambda N(t). \tag{2}$$

Eq. 1 is the solution to the rate equation, Eq. 2. The rate equation tells us that the number of radioactive atoms decaying per unit time at time t is λ times the number of radioactive atoms present at time t.

The half-life of the isotope is the time it takes for half of a sample of the isotope to decay. It is determined by setting

$$N(T_{1/2}) = 1/2 \, N(0) \tag{3}$$

or

$$N(0)e^{-\lambda T_{1/2}} = 1/2 \, N(0). \tag{4}$$

Canceling $N(0)$ and taking the $\log_e$ of both sides,

$$-\lambda T_{1/2} = \log_e 1/2 = -\log_e 2 \tag{5}$$

or

$$\lambda = \frac{\log_e 2}{T_{1/2}} = \frac{0.69}{T_{1/2}}. \tag{6}$$

For C^{14}, with a half-life of 5,700 years, $\lambda = 1/8,225$ years.

Wherever C^{14} atoms are, they will decay as just described. However, the production of C^{14} is very much dependent on location. In the atmosphere, production occurs as a result of cosmic radiation. Because cosmic rays do not penetrate seawater to any appreciable extent, C^{14} production can be assumed to be negligible in the oceans.[7] This absence of an internal production mechanism in seawater is the key to the lower C^{14}/C^{12} ratio in seawater than in the atmosphere. To turn this qualitative set of ideas into a quantitative result, we must make a mathematical model.

We will assume that the amounts of C^{14} and C^{12} in the atmosphere and in the ocean are constant, so that a steady-state box model can be used. Actually, because of relatively recent C^{14} production in the atmosphere from nuclear weapons testing and because of dilution of the atmospheric C^{14} concentration resulting from the release of $C^{12}O_2$ from fossil fuel burning,[8] the steady-state assumption is not a perfect one, but the corrections are small.

Next, we must construct an appropriate box model. You might be tempted to construct a model containing two boxes—atmosphere and

---

7. The same is true in biological materials. Whatever C^{14} a hunk of wood has it received while the tree was incorporating carbon from the atmosphere in photosynthesis. Once lodged in biological tissue, the C^{14} slowly decays away, but virtually no new C^{14} is formed.

8. Fossil fuels were formed many millions of years ago from biological materials. Since the half-life of C^{14} is 5,700 years, the original C^{14} in fossil fuel will have been reduced in half many times, so that now nearly none is left. Therefore, the CO_2 released to the atmosphere from fossil fuel burning dilutes the atmospheric C^{14} concentration, a process called the Suess effect.

ocean. However, an examination of their physical character suggests that the oceans themselves should each be thought of as a two-box system, with each box a horizontal layer. The upper, or mixed, layer is about 75 m thick and is relatively warm. The water in this layer is well stirred, and hence its name. The lower level is the deep ocean, which is colder and much larger in volume. Thermal stratification keeps these two "boxes" relatively isolated from one another, although some flow of water slowly mixes them. The boundary between the layers is called the thermocline, a transition zone that in some places is almost 1,000 m thick. As long as our interest is in the mixed layer and the deep layer, the details of what goes on in the transition zone need not concern us; and so this zone is ignored in our analysis.

Thus the simplest box model we can use to help us answer our problem is a three-box model, consisting of the atmosphere and the two oceanic layers.[9] Figure III-2 shows the major inflows and outflows of C^{12} and C^{14} for the three boxes. The amounts of C^{14} in the atmosphere and in dissolved form in the mixed layer and deep oceanic layer are denoted X_A, X_M, and X_D, respectively. The amounts of C^{12} in the atmosphere and in dissolved form in the oceanic layers are denoted Y_A, Y_M, and Y_D. The numerical values of the Y's are derivable from information in the Appendix (III, XII.5), but we will not require them to solve this part of the problem. C^{14} is produced in the atmosphere at a constant rate, R. The numerical value of R will not concern us—it is a constant determined by the cosmic ray flux. C^{14} can exit the atmosphere by decay at a rate λX_A (see Eq. 2). C^{14} can also pass across the atmosphere-ocean boundary by diffusion and across the thermocline in flowing water.

Diffusion processes are an example of a class called linear donor-controlled flows (see Problem II.7). The rate of diffusion of C^{14} from the atmosphere to the mixed oceanic layer is controlled solely by the amount of C^{14} in the atmosphere and is proportional to, or a linear function of, that amount. The proportionality constant is denoted by α in Figure III-2. Similarly, the rate of diffusion from the mixed oceanic layer to the atmosphere is proportional to the amount of C^{14} in the mixed layer. The proportionality constant is denoted by β. The net diffusion rate of C^{14} from the atmosphere to the mixed oceanic layer is $\alpha X_A - \beta X_M$. The constants α and β are not equal to each other; their ratio is related to Henry's constant, discussed in Chapter II Section C. In case you wonder why the values of α and β aren't given here or in the Appendix, relax; we can solve our problem without knowing in advance their actual values.

---

9. Broecker (1974) provides useful material on many small corrections left out in the treatment presented here. That superb text also contains a number of interesting applications of box models to other problems in chemical oceanography.

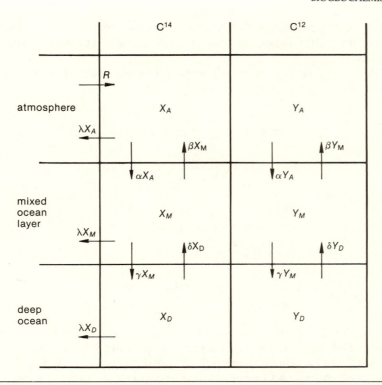

Figure III-2   The flows of C^{14} (X) and C^{12} (Y) among the atmosphere, the mixed ocean layer, and the deep ocean. R is the rate of formation of C^{14} and λ is its decay rate constant. α and β are diffusion rate constants, while γ and δ are convection rate constants.

The slow movement of water across the thermocline results in the exchange of C^{14} between the deep and mixed layers. This is also a linear, donor-controlled process. The rates are proportional to the amounts of C^{14}, with proportionality constants γ and δ as shown in Figure III–2.

$C^{12}O_2$ also flows between atmosphere and ocean and across the thermocline. Because C^{12} is neither created in the atmosphere nor lost from either box by radioactive decay, no other flows are included. We will take the proportionality constants for C^{12} diffusion across the atmosphere-ocean boundary and flow across the thermocline to be the same as for C^{14}. This is a very good approximation but it is not perfect. The small mass diffusion difference between $C^{14}O_2$ and $C^{12}O_2$ results in a slightly smaller constant for the heavier isotope.

There are other sources and sinks for carbon in each box as well: exchange with the biosphere, release from fossil fuel burning and forest fires, deposition to the ocean sediment, and release to the atmosphere by volcanoes. Even though one of these flows (exchange with the biosphere) is large, it does not affect our result (see Exercise 8). The other carbon flows are relatively small and affect our answer only slightly.

The steady-state condition is imposed by equating inflows and outflows of C^{12} and C^{14} for each box:

$$R + \beta X_M = \lambda X_A + \alpha X_A , \tag{7}$$

$$\alpha X_A + \delta X_D = \lambda X_M + \beta X_M + \gamma X_M , \tag{8}$$

$$\gamma X_M = \delta X_D + \lambda X_D , \tag{9}$$

$$\alpha Y_A = \beta Y_M , \tag{10}$$

and

$$\gamma Y_M = \delta Y_D. \tag{11}$$

Eqs. 7–9 tell us that the amounts of C^{14} in the atmosphere, in the mixed layer, and in the deep layer are constant; Eqs. 10 and 11 constrain C^{12} to be constant in each of the three boxes (see Exercise 1). Now we can proceed to answer the two questions posed in the problem statement.

(a) We want to determine the residence time of C^{12} in each of the oceanic layers. We are given the steady-state model expressed in Eqs. 7–11 and the empirical information that

$$\frac{\dfrac{X_D}{Y_D}}{\dfrac{X_A}{Y_A}} = 0.77 \tag{12}$$

and

$$\frac{\dfrac{X_D}{Y_D}}{\dfrac{X_M}{Y_M}} = 0.81 \tag{13}$$

The residence time of C^{12} in the deep layer, T_D^{12}, is, by definition,

$$T_D^{12} = \frac{Y_D}{\gamma Y_M} = \frac{1}{\delta} , \tag{14}$$

and the residence time of C^{12} in the mixed layer, T_M^{12}, is, by definition,

$$T_M^{12} = \frac{Y_M}{\alpha Y_A + \delta Y_D}. \qquad (15)$$

Our task is to express T_D^{12} and T_M^{12} in terms of quantities whose numerical values we know, such as the combinations of X's and Y's in Eqs. 12 and 13. Consider T_D^{12} first. Eq. 14 involves the unknown constant δ, but by use of Eqs. 9 and 11 (which are two equations containing two unknowns, δ and γ) we can solve for δ in terms of X's and Y's. Substituting the value of γ from Eq. 11 into Eq. 9, we obtain

$$\delta X_D = \frac{\delta Y_D}{Y_M} X_M - \lambda X_D \qquad (16)$$

or

$$\delta = \frac{\lambda X_D}{\dfrac{Y_D X_M}{Y_M} - X_D}. \qquad (17)$$

Substituting Eq. 17 into Eq. 14,

$$T_D^{12} = \frac{\dfrac{Y_D X_M}{Y_M} - X_D}{\lambda X_D}$$

$$= \frac{\dfrac{Y_D X_M}{Y_M X_D} - 1}{\lambda}. \qquad (18)$$

Using Eq. 13 and the value of λ for C^{14},

$$T_D^{12} = 1{,}930 \text{ yr.} \qquad (19)$$

To calculate the residence time of C^{12} in the mixed layer, T_M^{12}, we must eliminate the unknown constant α from Eq. 15. Using Eqs. 8–11, we can show that

$$\alpha = \frac{\lambda(X_M + X_D)}{X_A - \dfrac{Y_A X_M}{Y_M}}. \qquad (20)$$

If this and Eq. 17 are substituted into Eq. 15,

$$T_M^{12} = \frac{Y_M}{\dfrac{\lambda(X_M + X_D)Y_A}{X_A - \dfrac{Y_A X_M}{Y_M}} + \dfrac{\lambda X_D Y_D}{\dfrac{Y_D X_M}{Y_M} - X_D}}. \tag{21}$$

We want to express this, if possible, in terms of the combinations of X's and Y's in Eqs. 12 and 13. With a little algebraic maneuvering (see Exercise 2), we can show that Eq. 21 can be rewritten as

$$T_M^{12} = \frac{1}{\lambda \left[\dfrac{1 + \dfrac{X_D}{X_M}}{\dfrac{X_A Y_M}{Y_A X_M} - 1} + \dfrac{\dfrac{X_D}{X_M}}{1 - \dfrac{X_D Y_M}{Y_D X_M}} \right]}. \tag{22}$$

Note that $(X_A Y_M)/(Y_A X_M)$ can be written so that X_A/Y_A is the numerator and X_M/Y_M is the denominator. If we divide both the numerator and denominator of $(X_A/Y_A)/(X_M/Y_M)$ by X_D/Y_D, we get

$$\frac{X_A Y_M}{Y_A X_M} = \frac{\dfrac{X_A/Y_A}{X_D/Y_D}}{\dfrac{X_M/Y_M}{X_D/Y_D}} = \frac{\dfrac{1}{0.77}}{\dfrac{1}{0.81}} = 1.05. \tag{23}$$

This leaves only the ratio X_D/X_M in Eq. 22 to determine. Using Eq. 13, we find

$$\frac{X_D}{X_M} = 0.81 \frac{Y_D}{Y_M}. \tag{24}$$

If dissolved carbon is fairly uniformly distributed throughout the ocean, then the ratio Y_D/Y_M should roughly equal the ratio of the volume of the deep layer of the ocean to that of the mixed layer. For the Pacific Ocean, this ratio is about 40. Actually, the cold water of the deep ocean presently holds about a 15% higher concentration of dissolved carbon than does the warmer water of the mixed layer.[10] Therefore,

$$\frac{Y_D}{Y_M} = 1.15(40) = 46. \tag{25}$$

---

10. It is for the same reason that a warm beer goes flat: Warm beer cannot hold as much CO_2 in solution as does cold beer.

Now, using Eqs. 24 and 25, we get

$$\frac{X_D}{X_M} = 37. \tag{26}$$

Combining Eqs. 13, 22, 23 and 26, we calculate that

$$T_M^{12} = \frac{1}{\lambda\left[\dfrac{1+37}{1.05-1} + \dfrac{37}{1-0.81}\right]} \tag{27}$$
$$= 8.6 \text{ yr.}$$

We have now calculated the residence times of dissolved C^{12} in the two oceanic layers. Because the ratio of dissolved C^{14} to dissolved C^{12} in seawater is very small (about 10^{-12}) the C^{12} residence times are actually very nearly equal to the residence times of total dissolved carbon.

Knowing the residence time of dissolved inorganic carbon in the deep oceanic layer, you can estimate the residence time of deep ocean water itself. These residence times will be approximately equal, for the following reason: If the stock and flow of water are multiplied by the same quantity—the concentration of carbon—you obtain the stock and flow of carbon in that water. This reasoning is based on the valid assumption that diffusion of carbon across the thermocline is not significant compared to the movement of carbon across the thermocline in flowing water.

(b) The gross rate of exchange of carbon from the atmosphere to the ocean is given by $\alpha(Y_A + X_A)$. Because $X_A/Y_A << 1$, this can be approximated by αY_A. Using the stock-flow-residence time relation as expressed in Eq. 15, we find

$$\alpha Y_A = Y_M/T_M^{12} - \delta Y_D. \tag{28}$$

We now know T_M^{12} and δ (the latter can be obtained from Eqs. 14 and 19), so all we require are the values of Y_M and Y_D, the stocks of dissolved carbon in the mixed and deep layers. The concentration of dissolved inorganic carbon in the Pacific Ocean mixed layer is about 2.3×10^{-3} moles/liter, and the volume of the Pacific Ocean mixed layer is about 1.25×10^7 km³; a mole of carbon has a mass of 12 g. Therefore,

$$Y_M = 3.5 \times 10^{11} \text{ tonnes(C)}, \tag{29}$$

and the flow of carbon across the ocean-atmosphere interface is approximately

$$\alpha Y_A = \frac{Y_M}{T_M^{12}} - \delta Y_D$$

$$= Y_M \left[\frac{1}{T_M^{12}} - \gamma \right].$$

(30)

To determine γ, we use Eqs. 11 and 17, obtaining

$$\gamma = \frac{\delta Y_D}{Y_M}$$

$$= \frac{\lambda Y_D X_D}{Y_D X_M - X_D Y_M}$$

(31)

$$= \frac{\lambda \dfrac{Y_D}{Y_M}}{\dfrac{Y_D X_M}{Y_M X_D} - 1} = \frac{\lambda 46}{\dfrac{1}{0.81} - 1} = \frac{1}{41.94 \text{ yr}}.$$

Substituting Eqs. 27, 29, and 31 into Eq. 30, we obtain

$$\alpha Y_A = 3.5 \times 10^{11} \left(\frac{1}{8.6} - \frac{1}{41.9} \right) \text{tonnes(C)/yr}$$

$$= 3.2 \times 10^{10} \text{ tonnes(C)/yr}.$$

(32)

Note that the *net* flow of carbon between atmosphere and ocean is zero in our model[11]—what we've calculated here is the *gross* flow.

EXERCISE 1: Explain why two equations (10 and 11) suffice to constrain the amounts of C^{12} in all three boxes to be constant.

EXERCISE 2: Derive Eq. 20 from Eqs. 8–11 and derive Eq. 22 from Eq. 21.

EXERCISE 3: Calculate the residence time, T_D^{14}, of C^{14} in the deep layer of the Pacific Ocean by evaluating $X_D / \gamma X_M$.

EXERCISE 4: Calculate $(T_D^{12})^{-1} - (T_D^{14})^{-1}$ and interpret its value using your result from Exercise 3 of Problem II.6.

---

11. See Problem III.4 for a discussion of the more realistic case in which there is a net flow of CO_2 from atmosphere to ocean resulting from the disequilibrium of CO_2 produced from fossil fuel combustion.

EXERCISE 5:   Considering ocean uptake (by diffusion) and terrestrial biospheric uptake of atmospheric inorganic carbon (in net primary productivity; see Section XII.1 of the Appendix), what is the residence time of CO_2 in Earth's atmosphere?

EXERCISE 6:   What is the gross rate of flow of C^{12} between the deep ocean and the mixed layer?

* *EXERCISE 7:*   (*a*) Ignoring the flow of water in and out of the mixed layer by precipitation, runoff, and evaporation, what is the residence time of water in the mixed layer? (*b*) Using data from the Appendix and making a reasonable assumption about the fraction of oceanic evaporation that occurs from the Pacific Ocean, calculate the actual residence time of water in the mixed layer of the Pacific. Why is it not about equal to the residence time of mixed-layer dissolved inorganic carbon, despite the fact that the residence time of deep-layer seawater does roughly equal the residence time of deep-layer dissolved inorganic carbon?

* *EXERCISE 8:*   Suppose a fourth box is added to the system, containing terrestrial vegetation. Carbon now flows back and forth between the atmosphere and this fourth compartment. Explain why this will not affect the oceanic residence times for dissolved inorganic carbon.

4. Atmospheric CO_2 and the Ocean Sink

If a shot of CO_2 is added to the atmosphere, estimate how much of it will be removed eventually from the atmosphere and taken up by the oceans.

· · · · · · ·

Prior to the industrial revolution, the concentration of CO_2 in the atmosphere was about 275 ppm(v). Because the atmosphere contains 1.8×10^{20} moles of air, there were about $275 \times 10^{-6} \times 1.8 \times 10^{20}$ or 4.95×10^{16} moles of CO_2 in the atmosphere then. Presently (1983), the CO_2 concentration is about 340 ppm(v), corresponding to 6.12×10^{16} moles of CO_2—a gain of 1.17×10^{16} moles. If all the fossil fuel burned from 1800 to the present is added up and the total CO_2 emission estimated, it turns out that about 1.5×10^{16} moles of CO_2 were injected to the atmosphere throughout that period.[12] Therefore 1.17/1.5, or 78% of the CO_2 originally injected from burning fossil fuels is now present in the atmosphere. A likely possibility is that the oceans have taken up most, or all, of the other 22%. This problem explores that possibility.

As a rough guess we might assume that a shot of CO_2, injected initially into the atmosphere, will end up partitioned between the atmosphere and the oceans in proportion to the present number of moles of carbon in each of those carbon reservoirs. Provided all three species of dissolved inorganic carbon (H_2CO_3, HCO_3^-, and CO_3^{-2}) are included in the oceanic inventory, and provided the quantity of CO_2 initially injected is small compared to present stocks in the atmosphere and oceans, it seems reasonable to assume that the equilibrium ratio will remain constant.

Let's look at the implications of this assumption. Suppose, first, that all of Earth's oceans are a potential sink for excess atmospheric CO_2. There are about 55 times more moles of dissolved inorganic carbon in all of Earth's oceans than in the atmosphere. (This ratio can be derived from information in Appendix XIII.) In the eventual equilibrium there would then be a 1:55 partition of the new CO_2 between air and water. In other words, only 1/56 of the CO_2 would remain in the atmosphere and over 98% would be removed. Suppose, next, that only the mixed layer of the oceans is a sink for excess atmospheric CO_2. Then, since there is about as much dissolved inorganic carbon in the mixed layer of the oceans as there is CO_2 in the atmosphere, the equilibrium partition of new CO_2 between air and mixed layer

---

12. Other relatively small sources of CO_2, such as from cement manufacturing, are also included in this estimate (Clark 1982). A potentially large but highly uncertain source originating from deforestation is not included in this estimate but will be discussed below.

would, by the same reasoning, be in the ratio of about 1:1—one half of the injected CO_2 would be removed. Appealing as this simple argument is, however, it is wrong. The reactions governing the equilibrium concentrations of dissolved H_2CO_3, HCO_3^-, and CO_3^{-2} are sufficiently complex that our partitioning assumption is not valid.

To see what really happens, we have to look at the chemical reactions involved. In particular, as more CO_2 is dissolved in seawater, the net effect[13] is summarized by the following:

$$H_2O + CO_2 + CO_3^{-2} \rightarrow 2HCO_3^-. \tag{1}$$

Eqs. 8 and 9 in the introduction to Chapter II Section C tell us that in equilibrium,

$$[H^+][HCO_3^-] = K_1 [H_2CO_3] \tag{2}$$

and

$$[H^+][CO_3^{-2}] = K_2 [HCO_3^-], \tag{3}$$

where $K_i = 10^{-pK_i}$.

From the second of these relations, we determine that

$$[H^+] = \frac{K_2 [HCO_3^-]}{[CO_3^{-2}]}. \tag{4}$$

Hence, the process shown in Eq. 1, in which $[HCO_3^-]$ increases and $[CO_3^{-2}]$ decreases, leads to an increase in $[H^+]$. Eq. 2 informs us that as $[H^+]$ increases, so does the ratio

$$\frac{[H_2CO_3]}{[HCO_3^-]} = \frac{[H^+]}{K_1}. \tag{5}$$

Therefore, $[H_2CO_3]$ must increase even more than $[H^+]$, because the denominator, $[HCO_3^-]$, is increasing as well. According to Henry's law (Eq. 10 in the introduction to Chapter II, Section C), the concentration of CO_2 in the atmosphere in equilibrium will be proportional to $[H_2CO_3]$. Therefore, the increase in $[H_2CO_3]$ that results from the passage of CO_2 into seawater will inhibit the further passage of CO_2 from the atmosphere to the oceans. More than 1/56 of the new CO_2 will remain in the atmosphere.

To derive a quantitative result from this qualitative picture a slight detour is needed to explain the concept of alkalinity. Alkalinity is a

---

13. This net effect is a result of the reactions $CO_2 + H_2O \rightleftharpoons H_2CO_3$, $H_2CO_3 \rightleftharpoons H^+ + HCO_3^-$, and $HCO_3^- \rightleftharpoons H^+ + CO_3^{-2}$.

measure of the ability of a system to neutralize acid. Negatively charged chemical species that can combine with H^+ to form a less acidic product contribute to the alkalinity of a chemical system. In aqueous systems in which the only negatively charged, H^+-accepting, species are HCO_3^-, CO_3^{-2} and OH^-, the concentration of alkalinity, [Alk], is defined to be

$$[Alk] = [HCO_3^-] + 2[CO_3^{-2}] + [OH^-] - [H^+]. \qquad (6)$$

The reason for the factor of 2 in front of $[CO_3^{-2}]$ is that CO_3^{-2} is doubly charged and therefore 1 mole of CO_3^{-2} can neutralize 2 moles of H^+. Note that the concentration of H^+ is subtracted from the negatively charged terms in Eq. 6. Because other H^+-accepting species, such as $B(OH)_4^-$, are present in seawater, their concentrations should, in principle, be added to Eq. 6. To first approximation, however, such species can be ignored because their concentrations are relatively low.

In typical seawater, the alkalinity is about 2.4×10^{-3} moles per liter. The dominant contribution to alkalinity comes from HCO_3^-, with CO_3^{-2}, in second place, contributing about 10%. Because the pH of seawater is about 8.2, the $[H^+]$ and $[OH^-]$ terms in Eq. 6 are relatively unimportant, and hence, for seawater,

$$[Alk] = [HCO_3^-] + 2[CO_3^{-2}]. \qquad (7)$$

Consideration of Eq. 7 and the process described by Eq. 1 shows that the absorption of CO_2 in seawater will not change the alkalinity: Every mole of CO_3^{-2} that disappears as a result of the process in Eq. 1 is replaced by two new moles of HCO_3^-. This can only be true, however, if the supply of CO_3^{-2} is not depleted by this process. CO_3^{-2} is produced by the dissolution of $CaCO_3$, the stuff of seashells. Most of the oceans' waters are supersaturated with dissolved $CaCO_3$, so CO_3^{-2} can be readily removed from the water without causing the dissolution of more solid $CaCO_3$. In other words, the supply of $CaCO_3$ is large enough to insure a constant alkalinity for increases in atmospheric CO_2, at least for increases as large as would result from burning all the known fossil fuel resources on Earth.[14] Hence, we are justified in keeping the concentration of alkalinity constant in our examination of the partitioning of a relatively small shot of CO_2 to the atmosphere (such as the roughly 1.5×10^{16} moles of CO_2 that have been injected since the industrial revolution).

The quantity we want to calculate is the fraction, f, of carbon added initially to the atmosphere that remains there in the new equilibrium. We will let the symbol X denote the number of moles of CO_2

---

14. See Broecker (1974) for more discussion of this point.

present in the atmosphere. X_0 is the number of moles of CO_2 in the atmosphere prior to the injection. ΔX_0 is the amount of CO_2 initially injected to the atmosphere, and ΔX is the final equilibrated increase in atmospheric CO_2 after some portion of ΔX_0 has entered the oceans (see Figure III-3). Then

$$f = \frac{\Delta X}{\Delta X_0}. \tag{8}$$

Similarly, we use the symbol, Y, to denote the number of moles of dissolved inorganic carbon in seawater. If Y_0 equals the number of moles of dissolved inorganic carbon in the oceans prior to the injection of ΔX_0, and ΔY is the final equilibrated increase in dissolved inorganic carbon in ocean water as a result of the initial increase, ΔX_0, in atmospheric CO_2, then conservation of carbon leads to

$$\Delta X_0 = \Delta X + \Delta Y. \tag{9}$$

Hence

$$f = \frac{\Delta X}{\Delta X + \Delta Y}$$

$$= \frac{1}{1 + \dfrac{\Delta Y}{\Delta X}}. \tag{10}$$

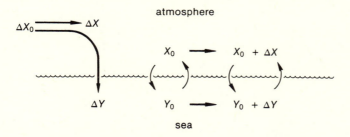

Figure III-3 The partition of new CO_2 between atmosphere and ocean. ΔX_0, the initial addition of CO_2 to the atmosphere, is partitioned into two parts. ΔX is the part that ultimately remains in the atmosphere and ΔY is the part that ends up in the ocean. X_0 and Y_0, the initial amounts of carbon in air and sea, are in equilibrium. Afterwards, $X_0 + \Delta X$ and $Y_0 + \Delta Y$ are again in equilibrium.

Our task,[15] then, is to derive an estimate for $\Delta Y / \Delta X$.

To make our calculation relevant to the major concern over the CO_2 greenhouse threat to Earth's climate we will assume that X_0 is the amount of CO_2 in the preindustrial atmosphere. The best current estimate is that the concentration of CO_2 then, $p(CO_2)_0$, was about 275 ppm(v), or

$$X_0 = 4.95 \times 10^{16} \text{ moles.} \tag{11}$$

Next, we need to know how much dissolved inorganic carbon was in the oceans at that time. We will assume that CO_2 exchange across the air-water boundary was in equilibrium then, so that

$$[H_2CO_3]_0 = K_H \, p(CO_2)_0, \tag{12}$$

where $[H_2CO_3]_0$ is the initial concentration of H_2CO_3 in the oceans and K_H is Henry's constant. The chemical reactions linking H_2CO_3, HCO_3^-, and CO_3^{-2} equilibrate very rapidly, so we can assume the additional equilibrium relations given by Eqs. 2 and 3 for the initial value of $[H^+]$, $[HCO_3^-]$, and $[CO_3^{-2}]$. Averaged over all the ocean waters, the equilibrium constants have the values[16]

$$K_H = 0.0382,$$
$$K_1 = 8.8 \times 10^{-7}, \tag{13}$$

and

$$K_2 = 5.6 \times 10^{-10}.$$

This information is not sufficient to calculate $[HCO_3^-]$, $[CO_3^{-2}]$, and $[H^+]$; an additional relation is needed. In Problem II.19, charge conservation provided that additional relation, but in seawater there are so many charged chemical species present that such an approach becomes more difficult than is necessary for the problem at hand.[17] For

---

15. Some authors work with the expression $(\Delta X / \Delta Y)(Y_0 / X_0)$, which is called the Revelle factor or the buffer factor.

16. These values differ from the ones used in Chapter II, Section C. There, in the discussion of acid precipitation, freshwater values were used. Here, we are dealing with seawater, which contains salts that alter the values of these constants (Stumm and Morgan 1981).

17. For a derivation starting from charge conservation and equilibrium relations of the approximate concentrations of major ionic species in seawater, see Stumm and Morgan (1981).

our purposes we will simply take an approximate value for preindustrial seawater pH as a given:[18]

$$[H^+]_0 = 10^{-8.3} \text{ moles/liter.} \tag{14}$$

Using Eqs. 2, 3, 7, 12, 13, and 14, along with the numerical value of 275 ppm(v) for $p(CO_2)_0$, we obtain

$$[H_2CO_3]_0 = 0.01 \times 10^{-3} \text{ moles/liter,} \tag{15}$$

$$[HCO_3^-]_0 = 1.85 \times 10^{-3} \text{ moles/liter,} \tag{16}$$

$$[CO_3^{-2}]_0 = 0.21 \times 10^{-3} \text{ moles/liter,} \tag{17}$$

and

$$\begin{aligned} [Alk]_0 &= [HCO_3^-]_0 + 2[CO_3^{-2}]_0 \\ &= 2.27 \times 10^{-3} \text{ moles/liter.} \end{aligned} \tag{18}$$

For the total concentration, $[C_T]_0$, of dissolved inorganic carbon, we obtain

$$\begin{aligned} [C_T]_0 &= [H_2CO_3]_0 + [HCO_3^-]_0 + [CO_3^{-2}]_0 \\ &= 2.07 \times 10^{-3} \text{ moles/liter.} \end{aligned} \tag{19}$$

The amount of inorganic carbon in the preindustrial ocean, Y_0, can now be derived by multiplying $[C_T]_0$ by the number of liters, M_s, of seawater. For the entire oceans, M_s equals 1.35×10^{21} liters, while for the mixed layer (assumed to be 75 m deep), $M_s = 2.7 \times 10^{19}$ liters. Therefore

$$Y_0^{whole\ ocean} = 279 \times 10^{16} \text{ moles} \tag{20}$$

18. This value of average seawater pH, 8.3, is about 2% greater than the correct value for deep ocean water. It is, in fact, closer to the average preindustrial surface water pH in the mixed layer. If a more accurate value is used, the value of [Alk] comes out too small by about 20%. The origin of these errors is in our neglect of the whole complex of ionic species present in real seawater. Even with a pH of 8.3, the concentrations of total dissolved inorganic carbon and alkalinity that result from our simplified assumptions are slightly low.

and

$$Y_0^{mixed\ layer} = 5.6 \times 10^{16} \text{ moles.} \tag{21}$$

These values are slightly lower than the true values, for the reason mentioned in footnote 18.

Now we must determine the values of ΔY and ΔX subsequent to an initial injection, ΔX_0, of CO_2 into the atmosphere. To do this we use the fact, discussed above, that [Alk] will remain nearly constant as $[C_T]$ increases. Our plan of attack is first to write the equilibrium value of Y as a fraction of the equilibrium value of X and of quantities that remain constant, like K_1, K_2, K_H, and [Alk]. Then we can take the derivative of Y with respect to X, and thus obtain an approximation to $\Delta Y/\Delta X$. This can then be substituted into Eq. 10 to determine f.

So, first we write

$$Y = M_s([H_2CO_3] + [HCO_3^-] + [CO_3^{-2}]). \tag{22}$$

Using the equilibrium relations, this can be rewritten as

$$Y = M_s\left(\frac{K_H X}{\beta}\right)\left(1 + \frac{K_1}{[H^+]} + \frac{K_1 K_2}{[H^+]^2}\right), \tag{23}$$

where β is the number of moles of air in the atmosphere (equal to 1.8×10^{20}). The next step is to write $[H^+]$ in terms of X, [Alk], and the equilibrium constants. This is achieved by noting that

$$
\begin{aligned}
[Alk] &= [HCO_3^-] + 2[CO_3^{-2}] \\
&= [H_2CO_3]\left(\frac{K_1}{[H^+]} + \frac{2K_1 K_2}{[H^+]^2}\right) \\
&= \left(\frac{K_H X}{\beta}\right)\left(\frac{K_1}{[H^+]} + \frac{2K_1 K_2}{[H^+]^2}\right).
\end{aligned}
\tag{24}
$$

Multiplying both sides of Eq. 24 by $[H^+]^2$, a quadratic equation for $[H^+]$ is obtained:

$$[H^+]^2 - X\lambda[H^+] - 2X\lambda K_2 = 0, \tag{25}$$

where

$$\lambda = \frac{K_H K_1}{\beta[Alk]} = 0.82 \times 10^{-25}. \tag{26}$$

Eq. 25 is solved to yield

$$[H^+] = \frac{\lambda X}{2}\left(1 \pm \sqrt{1 + \frac{8K_2}{\lambda X}}\right). \tag{27}$$

Only the $1 + (8K_2/\lambda X)^{1/2}$ solution is acceptable, as $[H^+]$ cannot be negative.

If Eq. 27 is substituted into Eq. 23, we obtain

$$Y = \frac{M_s K_H X}{\beta}\left[1 + \frac{2K_1}{\lambda X\left(1 + \sqrt{1 + \frac{8K_2}{\lambda X}}\right)} + \frac{4K_1 K_2}{\lambda^2 X^2\left(1 + \sqrt{1 + \frac{8K_2}{\lambda X}}\right)^2}\right], \tag{28}$$

giving us the desired expression—a relation between only two variables, Y and X, with all other parameters in the equation equal to constants.

Differentiating Eq. 28 yields

$$\frac{dY}{dX} = \frac{M_s K_H}{\beta}\left[1 + \frac{8K_1 K_2}{\lambda^2 X^2\left(1 + \sqrt{1 + \frac{8K_2}{\lambda X}}\right)^2 \sqrt{1 + \frac{8K_2}{\lambda X}}} - \frac{4K_1 K_2}{\lambda^2 X^2\left(1 + \sqrt{1 + \frac{8K_2}{\lambda X}}\right)^2} + \frac{32K_1 K_2^2}{\lambda^3 X^3\left(1 + \sqrt{1 + \frac{8K_2}{\lambda X}}\right)^3 \sqrt{1 + \frac{8K_2}{\lambda X}}}\right]. \tag{29}$$

This expression can be evaluated at any value of X. If $X = X_0$ is substituted in, along with numerical values for K_H, K_1, K_2, λ, and β, we should obtain a value for dY/dX approximate to preindustrial conditions. Doing so, we obtain

$$\left.\frac{dY}{dX}\right|_{X = X_0} = 3.1 \times 10^{-21} M_s. \tag{30}$$

Now, M_s can be taken to be either the whole-ocean value or the mixed-layer value. If the former is used, then the f that is obtained when Eq. 30 is substituted into Eq. 10 will describe the equilibrium partitioning resulting from equilibrium with the entire ocean. Since the time scale for that equilibrium is, from Problem III.3, on the order of a millenium, that number will not be too relevant for understanding present partitioning of CO_2 released to the atmosphere within the past century or two. In other words, equilibrium throughout the whole ocean will not yet have been achieved. Nevertheless, it is interesting to see what value is obtained if $M_s = 1.35 \times 10^{21}$ liters is used. The result is

$$\left. \frac{dY}{dX} \right|_{X = X_{0,\ whole\ ocean}} = 4.2 \tag{31}$$

or, from Eq. 10,

$$\left. f \right|_{X = X_{0,\ whole\ ocean}} = \frac{1}{1 + 4.2} = 0.19. \tag{32}$$

Thus, 81% of the CO_2 added initially to the atmosphere will be taken up by the entire ocean after equilibrium is reached.

If equilibrium with only the mixed-layer is assumed and uptake of carbon by the deep ocean is neglected, then

$$\left. \frac{dY}{dX} \right|_{X = X_{0,\ mixed\ layer}} = 0.084 \tag{33}$$

or, from Eq. 10,

$$\left. f \right|_{X = X_{0,\ mixed\ layer}} = \frac{1}{1 + 0.084} = 0.92. \tag{34}$$

Thus, 8% of the CO_2 will be taken up by the mixed layer of the oceans after equilibration is reached.

In reality, about 22% of the CO_2 added to the atmosphere since the industrial revolution has been removed to the oceans.[19] This is compatible with Eqs. 10 and 30 if M_s is given by

$$f = \frac{1}{1 + M_s(3 \times 10^{-21})} = 0.78 \tag{35}$$

or

$$M_s = 0.91 \times 10^{20} \text{ liters.} \tag{36}$$

This corresponds to a depth of seawater equal to about 250 m. Of course, the new CO_2 in the ocean is not confined to the top 250 m; this is an effective depth, indicating the volume of water required to hold the CO_2 in equilibrium with atmospheric CO_2.

The average length of time since all the CO_2 molecules from fossil fuel burning were added to the atmosphere is about 25 yr.[20] Ocean circulation models suggest that in a period of 25 yr a quantity of CO_2 corresponding to an effective depth of 250 m could, indeed, have been absorbed in the oceans. So, if this were the whole picture, our understanding of the fate of CO_2 emissions to the atmosphere would be in good shape. However, there are two pieces of information that cause scientists to be doubtful that all is clear. First, deforestation probably has been a large contributor of atmospheric CO_2. Woodwell et al. (1983) estimate that from the year 1860 to 1980, between 1.1 and 1.9×10^{16} moles of CO_2 were lost from the biosphere to the atmosphere from global deforestation. Suppose the true value were 1.5×10^{16} moles of CO_2. Then 39% of all the CO_2 added to the atmosphere from both sources (fossil fuels and deforestation) now remains in the atmosphere. This implies

$$\frac{1}{1 + M_s(3.1 \times 10^{-21})} = 0.39 \tag{37}$$

---

19. Repeating an earlier point, this assumes that deforestation was not a major source of additional CO_2. It also assumes that the biosphere has not been a sink for extra atmospheric CO_2. If deforestation has been a major net source of CO_2, then a larger fraction than 22% of what was added has been removed. On the other hand, if the biosphere has been a net sink for CO_2, then less than 22% has entered the oceans. Our discussion below assumes that the removal of CO_2 from the atmosphere occurred by ocean uptake mediated by inorganic carbon chemistry in the sea and that the alterations of the biosphere had no net impact on the CO_2 budget.
20. This is much less than half the time elapsed since the beginning of fossil fuel combustion because combustion has accelerated enormously in recent years (see Exercise 4).

or

$$M_s = 5.0 \times 10^{20} \text{ liters.} \tag{38}$$

This corresponds to an effective depth of 1,400 m. Oceanographers say this is too deep; their ocean circulation models predict that no more than about half this much CO_2 could have been absorbed in the oceans in the time available.

The second piece of evidence indicating that our understanding of the fate of CO_2 is not complete comes from a series of measurements of atmospheric CO_2 concentrations in Hawaii from 1958 to the present. These data (Keeling et al. 1982) suggest that during this period, about 60% of the amount of the CO_2 added to the atmosphere from fossil fuel burning, alone, now remains there. Again, this upsets the oceanographers. Perhaps the oceanographic models are incorrect, or perhaps the biosphere has recently been a net sink, not a source, of CO_2. Resolution of this puzzle is an active area of current research.

EXERCISE 1:   Derive Eq. 29 from Eq. 28 by differentiating, and show by direct substitution of appropriate numerical quantities into Eq. 29 that Eq. 30 follows.

* *EXERCISE 2:*   Evaluate dY/dX at the present value of X rather than at X_0, both for the mixed layer and the whole ocean. Does your result suggest that we should be more or less worried about the future rate of buildup of CO_2 in the atmosphere if we use the present value of X rather than the preindustrial one?

EXERCISE 3:   Calculate the change in the pH of seawater from our assumed preindustrial value of 8.3 to its eventual whole-ocean equilibrium value, given that the atmosphere equilibrates (several millennia after fossil fuel burning ends) at 400 ppm(v) of CO_2. Although the absolute value of the pH you calculate here is not accurate, this estimate of the change in pH is fairly reliable.

* *EXERCISE 4:*   Suppose the rate of worldwide fossil fuel consumption has been growing at the rate of $a\%$ per year from $t = 0$ to the present, $t = T_0$. Derive an expression for the average "age" of the CO_2 emitted during this period (that is, the average of the time intervals from when each CO_2 molecule was released to the present). Your answer should depend only on a and T_0.

* *EXERCISE 5:*   Consider a container of pure fresh water in equilibrium with atmospheric CO_2. Show that the alkalinity of this aqueous system is zero, using the definition of alkalinity in Eq. 6. Next, assume a lump of limestone is placed in the water and it partially dissolves to produce a saturated solution. When this happens, Eq. 13 in the introduction to Chapter II, Section C is applicable. Calculate the alkalinity of this solution in units of micromoles per liter.

** *EXERCISE 6:* Consider a lake in equilibrium with atmospheric CO_2 and with an initial quantity of dissolved calcium carbonate to produce an alkalinity of 10^{-4} moles per liter. What is the initial pH of that solution? Now assume that HNO_3 is slowly added. The alkalinity will buffer the solution—that is, it will cause the pH to drop much more slowly than if there were no alkalinity present. In the process of buffering the acid, however, the alkalinity will be consumed. Using the methods described in Chapter II, Section C, demonstrate the existence of this buffer effect by calculating and plotting graphs of the pH and the alkalinity of the solution as a function of the amount of acid added. Assume the lakewater is a "closed system" in the sense that no new alkalinity is added to the lake or is generated in the lake sediments. You will find that as acid is added, the pH of the solution drops relatively little, whereas the alkalinity is reduced in proportion to the amount of acid added. Only when the alkalinity is nearly consumed by the acid will the pH begin to drop sharply (see Figure III-4). Alkalinity is a good measure of a lake's buffering capacity (i.e., ability to protect itself against acid precipitation). Lakes that have low alkalinity (for example, less than a few tens of micromoles per liter) such as many lakes in Scandinavia, the Adirondacks, eastern Canada, and the California Sierra Nevada, are particularly vulnerable to acid precipitation. Indeed, many lakes in the Adirondacks and Scandinavia have already lost all their alkalinity and suffered the sharp pH drop shown in Figure III-4. We say that such lakes are acidified.

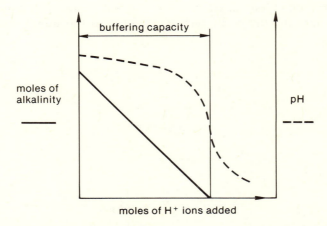

Figure III-4 The effect on pH (right-hand axis) and alkalinity (left-hand axis) of adding acid to a closed aqueous system. As more acid is added, the alkalinity steadily declines, whereas pH declines only gradually until the alkalinity is used up; at that point, pH dips sharply. Every mole of H^+ added to the system is neutralized by a mole of alkalinity until the alkalinity is used up in the process. The alkalinity present initially is thus a measure of the lake's buffering capacity—its ability to withstand acid input without suffering a large drop in pH.

5. A Perturbed Phosphorus Cycle (II)

Suppose some inorganic phosphorus is suddenly added to a lake in which phosphorus flows in a closed cycle. If the amount added is relatively small compared to the amount already present, how will the amount of phosphorus in each major compartment of the lake (inorganic phosphorus, phosphorus in dead organic matter, and phosphorus in living biomass) behave after this perturbation?

· · · · · · ·

Solving this problem requires the use of matrix methods and the introduction of some mathematics that will be unfamiliar to many readers. However, the key formula used in the solution, Eq. 13, is so useful in the quantitative analysis of interconnected systems that the courageous reader is urged to persevere.[21]

The phosphorus model used in Problem II.8 can be used to explore this question. The reader is urged to go back to that problem and recall how the model was structured. We assume the same initial conditions here as in Problem II.8, so that prior to the perturbation $X_1 = 0.2$, $X_2 = 0.1$, $X_3 = 1.0$ (in units of micromoles(P) per liter), and, at $t = 0$, X_2 is increased to a value of 0.12. Because we are now interested in the behavior in time of the perturbed system, we need to write a differential equation for each $X_i(t)$. The equations will simply state that the rate of change of each X_i is equal to the inflow minus the outflow:

$$\frac{dX_1}{dt} = \beta X_2 X_1 - \gamma X_1, \tag{1}$$

$$\frac{dX_2}{dt} = \alpha X_3 - \beta X_2 X_1, \tag{2}$$

and

$$\frac{dX_3}{dt} = \gamma X_1 - \alpha X_3. \tag{3}$$

---

21. Only one other problem in the book requires this mathematical technique (Exercise 7 of Problem III.5), so readers who are not familiar with matrix manipulations can skip over the remainder of this problem and continue on. The treatment that follows avoids derivations and merely presents results. Boyce and DiPrima (1973) have written an excellent reference for derivations of the mathematical formalism presented in a cursory manner here. Swartz (1973) reviews matrix algebra.

As in Eqs. 8–10 in Problem II.8, $\gamma = 0.25$, $\beta = 2.5$, and $\alpha = 0.05$ in time units of days. Note that

$$\frac{d}{dt} (X_1 + X_2 + X_3) = 0. \tag{4}$$

This derivative vanishes identically—that is, it vanishes for *any* values of the X's and the rate constants α, β, and γ. Integration of Eq. 4 gives

$$X_1(t) + X_2(t) + X_3(t) = \text{constant}, \tag{5}$$

implying that the total amount of phosphorus in the system is conserved—it doesn't leak away or pile up spontaneously. This is a consequence of the fact that phosphorus was assumed to flow in a closed cycle in the lake.

Our task is to solve Eqs. 1–3, or at least find a reasonable approximate solution, under the assumption that $X_1(0) = 0.2$, $X_2(0) = 0.1 + 0.02$, and $X_3(0) = 1.0$. If the solutions to Eqs. 1–3 approach a new steady state as time progresses, it must be the steady state we found in Problem II.8; namely, $\overline{X}'_1 = 0.203$, $\overline{X}'_2 = 0.1$ (unchanged from $\overline{X}_2$), and $\overline{X}'_3 = 1.017$. There are two possibilities: either the solutions to Eqs. 1–3 eventually settle down at these new steady-state values or they never settle down to a steady state. If the first is the case, then if we follow the solutions, $X_i(t)$, through a long enough period, we ought to see the $X_i(t)$ approach these values of $\overline{X}'_i$. Mathematically, we would say that as t approaches ∞, the $X_i(t)$ approach the $\overline{X}'_i$.

Eqs. 1–3 are examples of coupled, nonlinear differential equations. They are coupled in the sense that X_1 affects dX_2/dt, etc. They are nonlinear in the sense that the derivatives depend on products, not just sums, of the X_i. Such equations generally cannot be solved exactly; an approximation method is needed. The approximation method we use here is based on the fact that the initial perturbation is relatively small: The added phosphorus, 0.02 μm/liter, is small compared to the $\overline{X}_i$. Interestingly, we will see below that our method of solution provides a very general procedure for determining whether or not a complex system settles down to a new steady state after a small disturbance.

We will introduce the approximation method in a very general context and then apply it to our special case (Eqs. 1–3). We first have to introduce the concepts of eigenvalue and eigenvector. Given an $N \times N$ matrix, **A,**

$$\mathbf{A} = \begin{bmatrix} a_{11} & a_{12} & a_{13} & \cdot & \cdot & \cdot & a_{1N} \\ a_{21} & a_{22} & a_{23} & \cdot & \cdot & \cdot & a_{2N} \\ a_{31} & a_{32} & a_{33} & \cdot & \cdot & \cdot & a_{3N} \\ \cdot & \cdot & \cdot & & \cdot & \cdot & \cdot \\ \cdot & \cdot & \cdot & & \cdot & \cdot & \cdot \\ a_{N1} & a_{N2} & a_{N3} & \cdot & \cdot & \cdot & a_{NN} \end{bmatrix}, \tag{6}$$

the eigenvalues $\lambda_\sigma(\sigma = 1, 2, \ldots, N)$ are found by solving the equation

determinant $(\mathbf{A} - \lambda I) =$

$$\text{determinant} \begin{bmatrix} a_{11} - \lambda & a_{12} & a_{13} & \cdot & \cdot & \cdot & a_{1N} \\ a_{21} & a_{22} - \lambda & a_{23} & \cdot & \cdot & \cdot & a_{2N} \\ a_{31} & a_{32} & a_{33} - \lambda & \cdot & \cdot & \cdot & a_{3N} \\ \cdot & \cdot & \cdot & & \cdot & \cdot & \cdot \\ \cdot & \cdot & \cdot & & \cdot & \cdot & \cdot \\ a_{N1} & a_{N2} & a_{N3} & \cdot & \cdot & \cdot & a_{NN} - \lambda \end{bmatrix} = 0. \tag{7}$$

If this determinant is written out, the resulting expression will be an Nth order polynomial for the λ's, with coefficients that depend on the values of the a_{ij}. Therefore, Eq. 7 will have N solutions, $\lambda_1, \ldots \lambda_N$.

Given the eigenvalues, the eigenvectors, $\mathbf{U}^\sigma$ ($\sigma = 1, \ldots, N$) are determined by the N equations

$$\sum_{j=1}^{N} \mathbf{A}_{ij} \mathbf{U}_j^\sigma = \lambda_\sigma \mathbf{U}_i^\sigma. \tag{8}$$

Each $\mathbf{U}^\sigma$ is an N-dimensional vector, which is often represented as a column

$$\mathbf{U}^\sigma = \begin{bmatrix} \mathbf{U}_1^\sigma \\ \mathbf{U}_2^\sigma \\ \cdot \\ \cdot \\ \cdot \\ \mathbf{U}_N^\sigma \end{bmatrix}. \tag{9}$$

Now let's return to our problem, but in a more general form. Consider a system of equations (like Eqs. 1–3) for the time derivatives of a set of variables $(X_1, \ldots, X_N)$:

$$\frac{dX_i}{dt} = F_i(X_1, X_2, \ldots, X_N; a_1, \ldots, a_m). \tag{10}$$

The a_i are parameters, such as α, β, and γ in Eqs. 1–3, upon which the time derivatives depend. The steady-state conditions for the unperturbed X_i are determined by setting the time derivatives equal to zero and solving the resulting algebraic equations for the X:

$$\left. \frac{dX_i}{dt} \right|_{\substack{X_1 = \overline{X}_1 \\ \cdot \\ \cdot \\ \cdot \\ X_N = \overline{X}_N}} = F_i(\overline{X}_1, \overline{X}_2, \ldots \overline{X}_N; a_1, \ldots, a_m) = 0. \tag{11}$$

After a perturbation in which some of the X_i are displaced from their steady-state values, $\overline{X}_i$, the system will no longer be in a steady state, at least temporarily. The behavior of our system in the postperturbation period ($t > 0$ in our problem and in what follows) must satisfy Eq. 10 because that set of equations governs the system at *any* time. We denote the differences between the perturbed solution and the preperturbation solution by $Y_i(t)$, or in equation form:

$$Y_i(t) = X_i(t) - \overline{X}_i. \tag{12}$$

If the Y_i are much smaller in absolute value than the $\overline{X}_i$, it can be shown that subsequent to the perturbation at $t = 0$,

$$Y_i(t) = \sum_{j=1}^{N} \sum_{\sigma=1}^{N} \mathbf{C}_{i\sigma}(\mathbf{C}^{-1})_{\sigma j} Y_j(0) \, e^{\lambda_\sigma t}. \tag{13}$$

The λ_σ in Eq. 13 are the eigenvalues of an $N \times N$ matrix,[22] $\mathbf{M}$, whose matrix elements are given by derivatives of the functions, F_i, in Eq. 10:

$$\mathbf{M}_{ij} = \left. \frac{\partial F_i}{\partial X_j} \right|_{X_k = \overline{X}_k}. \tag{14}$$

The notation in Eq. 14 means that the partial derivatives are to be evaluated at $X_k = \overline{X}_k$ for all k. The matrix $\mathbf{C}$ in Eq. 13 is an $N \times N$

---

22. The matrix, $\mathbf{M}$, is called a community matrix in theoretical ecology. As initially introduced in ecology, the community matrix described the interaction of species; X_i, in other words, was the population size of the ith species. Subsequently, the meaning was broadened to include cases in which the X's stood for whole groups of organisms (for example, herbivores or carnivores) and even for nonliving components of ecosystems such as chemical nutrients. In a delightful and clear manner, May (1973) discusses application of the community matrix to stability problems in theoretical ecology.

matrix formed by taking the N eigenvectors of $\mathbf{M}$ and arranging them as columns. That is,

$$\mathbf{C} = [\mathbf{U}^1 \, \mathbf{U}^2 \, \mathbf{U}^3 \ldots \mathbf{U}^N], \tag{15}$$

where each $\mathbf{U}^\sigma$ is an N-dimensional eigenvector associated with the eigenvector, λ_σ. The initial perturbation is $Y_i(0)$. The smaller the Y_i are in comparison with the X_i, the more reliable is Eq. 13. In our problem

$$Y_1(0) = X_1(0) - \overline{X}_1 = 0, \tag{16}$$

because X_1 is living biomass and no phosphorus was added to that box at $t = 0$. However,

$$Y_2(0) = X_2(0) - \overline{X}_2 = 0.02, \tag{17}$$

while

$$Y_3(0) = X(0) - \overline{X}_3 = 0. \tag{18}$$

With this mathematical apparatus in hand, the solution to our problem is straightforward. The community matrix is determined by applying the definition of $\mathbf{M}$ (Eqs. 11 and 14) to Eqs. 1–3 with the values of α, β, and γ given by Eqs. 8, 9, and 10 of Problem II.8. The result is

$$\mathbf{M} = \begin{bmatrix} 0 & 0.5 & 0 \\ -0.25 & -0.5 & 0.05 \\ 0.25 & 0 & -0.05 \end{bmatrix}. \tag{19}$$

Using Eq. 7, the eigenvalues can be determined. The polynomial equation for the λ_σ is

$$\lambda^3 + 0.55\lambda^2 + 0.15\lambda = 0. \tag{20}$$

This cubic equation can be factored into

$$\lambda(\lambda^2 + 0.55\lambda + 0.15) = 0, \tag{21}$$

and so the three eigenvalues are

$$\lambda_1 = 0$$

$$\lambda_2 = \frac{-0.55 + 0.2975i}{2} \tag{22}$$

$$\lambda_3 = \frac{-0.55 + 0.2975i}{2}.$$

The eigenvectors are easily shown to be[23]

$$
\mathbf{U}^1 = \begin{vmatrix} 1 \\ 0 \\ 5 \end{vmatrix}, \; \mathbf{U}^2 = \begin{vmatrix} 1 \\ 2\lambda_2 \\ (0.2 + 4\lambda_2)^{-1} \end{vmatrix},
$$

$$
\mathbf{U}^3 = \begin{vmatrix} 1 \\ 2\lambda_3 \\ (0.2 + 4\lambda_3)^{-1} \end{vmatrix}. \tag{23}
$$

The matrix, **C**, is

$$
\mathbf{C} = \begin{bmatrix} 1 & 1 & 1 \\ 0 & 2\lambda_2 & 2\lambda_3 \\ 5 & (0.2 + 4\lambda_2)^{-1} & (0.2 + 4\lambda_3)^{-1} \end{bmatrix} \tag{24}
$$

and the matrix $\mathbf{C}^{-1}$, in Eq. 13, is

$$
\mathbf{C}^{-1} = \frac{\begin{bmatrix} 2\lambda_2(0.2 + 4\lambda_3)^{-1} & (0.2 + 4\lambda_2^{-1} & 2\lambda_3 \\ -\ 2\lambda_3(0.2 + 4\lambda_2)^{-1} & -(0.2 + 4\lambda_3)^{-1} & -2\lambda_2 \\ 10\lambda_3 & (0.2 + 4\lambda_3)^{-1} - 5 & -2\lambda_3 \\ -10\lambda_2 & -(0.2 + 4\lambda_2)^{-1} + 5 & 2\lambda_2 \end{bmatrix}}{\text{determinant } (\mathbf{C})}. \tag{25}
$$

Hence, from Eq. 13, we find

$$
\begin{aligned}
X_1(t) &= \overline{X}_1 + Y_1\,(t) \\
&= \overline{X}_1 + \frac{0.02}{\text{determinant } (\mathbf{C})} \\
&\quad \times \{e^{\lambda_1 t}[(0.2 + 4\lambda_2)^{-1} - (0.2 + 4\lambda_3)^{-1}] \\
&\quad + e^{\lambda_2 t}[(0.2 + 4\lambda_3)^{-1} - 5] \\
&\quad + e^{\lambda_3 t}[-(0.2 + 4\lambda_2)^{-1} + 5]\}
\end{aligned} \tag{26}
$$

$$
\begin{aligned}
X_2(t) &= \overline{X}_2 + \frac{0.02}{\text{determinant } (\mathbf{C})} \{e^{\lambda_1 t}[0] \\
&\quad + e^{\lambda_2 t}[2\lambda_2][(0.2 + 4\lambda_3)^{-1} - 5] \\
&\quad + e^{\lambda_3 t}[2\lambda_3][-(0.2 + 4\lambda_2)^{-1} + 5]\},
\end{aligned} \tag{27}
$$

---

23. Only the ratios of the elements of the eigenvalues are determined by Eq. 8. An overall normalization constant is arbitrary and will not affect the numerical value of the Y_i calculated from Eq. 13.

and

$$X_3(t) = \overline{X}_3 + \frac{0.02}{\text{determinant } (\mathbf{C})}$$
$$\times \{e^{\lambda_1 t}[5][(0.2 + 4\lambda_2)^{-1} - (0.2 + 4\lambda_3)^{-1}]$$
$$+ e^{\lambda_2 t}[0.2 + 4\lambda_2]^{-1}[(0.2 + 4\lambda_3)^{-1} - 5] \quad (28)$$
$$+ e^{\lambda_3 t}[0.2 + 4\lambda_3]^{-1}[-(0.2 + 4\lambda_3)^{-1} + 5]\}$$

It is reassuring to check that the $X_i(t)$ given by Eqs. 26–28 have the correct values at certain extreme, or limiting, values of t. If, for example, there were no initial change in the X_i, then each $X_i(t)$ should equal $\overline{X}_i$. This is indeed the case; the initial change (in X_2) was 0.02; if that term is set equal to zero in Eqs. 26–28, you can see that each $X_i(t)$ does equal $\overline{X}_i$.

A different limit of interest is that of large t. As t approaches infinity, the values of the $X_i(t)$ are also easy to deduce. First note that the term $e^{\lambda_1 t}$ in Eqs. 26–28 is a constant because $\lambda_1 = 0$. The only time dependence comes from $e^{\lambda_2 t}$ and $e^{\lambda_3 t}$. Because λ_2 and λ_3 have negative real parts[24] ($-0.55/2$), $e^{\lambda_2 t}$ and $e^{\lambda_3 t}$ both go to zero as t becomes large. We say that these terms "damp out." Thus, to calculate the values of $X_i(t)$, as $t \to \infty$, you simply ignore the $e^{\lambda_2 t}$ and $e^{\lambda_3 t}$ terms in Eqs. 26–28 (see Exercise 2).

This simple behavior at large t resulted from the fact that the real parts of the eigenvalues were negative. Had the real parts of any of the eigenvalues been positive, the exponential term, $e^{\lambda t}$, would have "blown up" as t became large and no new steady state would have been reached.[25]

EXERCISE 1: Derive Eq. 19.

EXERCISE 2: Show that Eqs. 26–28 are consistent with Eqs. 13, 17, and 18 of Problem II.8 by taking the $t \to \infty$ limit of Eqs. 26–28.

* *EXERCISE 3:* Derive Eqs. 23–25, consulting Swartz (1973) or any text on linear algebra if you need to look up the procedure for inverting a matrix.

***EXERCISE 4:* Assume a sudden 10% reduction occurs in the constant α in the phosphorus model. Approximately how will the X_i behave after the perturbation? [Hint: To the three system equations (Eqs. 1–3) tack on a fourth: $d\alpha/dt = 0$. This is purely an artifice, but

---

24. Swartz (1973) provides a good review of complex numbers.
25. Looking at the real parts of eigenvalues is a major tool for studying the stability of complex systems. May (1973) discusses this in more detail in his exceptionally lucid book. See also Boyce and DiPrima (1973).

it allows α to be treated as a system variable to be perturbed just as the X_i were before. The constraint, $d\alpha/dt = 0$, guarantees that once you perturb it, say by a 10% decrease, it will stay permanently at that new value. Now **M** will be a 4×4 matrix, and the initial conditions in Eq. 13 will be $Y_1(0) = Y_2(0) = Y_3(0) = 0$, and $Y_4(0) = \alpha(0) - \bar{\alpha} = -0.1\bar{\alpha}$, with $\bar{\alpha}$ given by Eq. 10 in Problem II.8.]

B.
Climatology

The first three problems in this subsection involve global climatic phenomena. In the first, Problem III.6, a model of some globally averaged features of Earth's present climate is developed. A few physical principles that govern the interaction of matter and energy, and the results of a few measurements, are the mortar and bricks of the model.

With a reasonable model of present climate in hand, we can then tinker on paper with the quantities that influence the climate. This is important to do, because the human population is in effect tinkering with these same quantities on a regional or global scale as a result of such activities as fossil fuel burning. With the aid of models, the kinds of climatic catastrophes human beings might trigger can be explored. Problems III.7 and 8 do just that.

Not all potential climatic alterations, however, occur on a regional or global scale. In densely populated urban areas, human beings are even now living to some extent in climates of their own making. Problem III.9 explores one important facet of this.

There are a great many theories about the natural causes of climatic change and about how human activity may be modifying Earth's future climate. Climate is influenced by numerous factors including the intensity of the sun's radiation; the geometry of Earth's orbit around the sun; the tilt of Earth's axis; the amount of water vapor, clouds, carbon dioxide, dust, and other substances in the atmosphere; the ratio of land to sea area, the location of the continents; and the types of covering, such as vegetation, on the continental surfaces. Phenomena such as volcanic eruptions, deforestation, fossil fuel burning, and nuclear war can alter some of these factors and thereby alter climate.

In the field of climatology it is not generally possible to ascribe a single cause to an observed effect; agreement over even the main causes of past trends in Earth's climate is lacking. The aim here is not

158

to resolve these major uncertainties. Rather, it is to develop a picture of the global characteristics of our climate and to show you how to estimate the major climatic consequences of some of the climate-altering activities of our species.

6. Earth's Surface Temperature

What is the globally averaged surface temperature of Earth?

· · · · · · ·

In Problem II.13, the blackbody temperature of Earth was derived by using the principle of energy conservation. The derivation involved a radiation balance equation in which the flux of outgoing infrared radiation, σT^4, was equated to the nonreflected portion of the solar flux, $\Omega(1-a)/4$. For that calculation, the only radiation transfer considered was that between the earth-atmosphere system and the sun or outer space.

To calculate the surface temperature, it is necessary to consider radiation transfer (a) between Earth's surface and atmosphere, and (b) within the atmosphere. A comparison of Earth and Venus sheds light on this. Both planets have about the same blackbody temperature; Earth's is about 250 K and Venus's is about 230 K. Yet Venus has a surface temperature of about 750 K, which is considerably hotter than Earth's globally averaged value of 290 K. What causes this difference is the much thicker, infrared-absorbing atmosphere on Venus. When infrared radiation (IR) is absorbed by gases in the atmosphere, reradiated, and subsequently absorbed by materials on a planet's surface, the surface is warmed. Absorption and reradiation of the infrared will occur many times in a thick atmosphere, increasing the warming. On a planet with a thin atmosphere, IR produced when the solar flux strikes planetary molecules has a low probability of being absorbed in the atmosphere and a higher probability of escaping to outer space before it can warm the surface.

How can this qualitative explanation of surface warming be converted into a quantitative model? Global climate models fall into two broad categories. First are the one-dimensional models, which describe only the vertical dimension. In such models, latitudinal and longitudinal variations of climatic parameters, such as temperature or pressure, are "averaged over" (in other words, just plain ignored). The second category of models takes into account horizontal variations. Because one-dimensional models are much easier to analyze, we will begin with one. Notice, though, that such models cannot provide any insight into certain issues, such as how the temperature change due to an increase in atmospheric CO_2 is distributed from equator to pole, or how poleward convection of air influences the globally averaged temperature change.

There is another way of classifying global climate models. In the most sophisticated models, the atmosphere is treated as a continuous medium; the equations describing flow of air and heat are differential or integro-differential equations with both space and time as independent variables. In other models, the atmosphere is divided into boxes or zones, and uniformity of conditions within each zone is assumed. In such models, the equations describing flows of heat or of air molecules usually describe time as a continuous variable, but spatial variations are described by algebraic equations rather than by differential calculus. Some models combine the two approaches, treating the vertical dimension as a continuum but dividing the horizontal dimension into discrete zones.

We'll use the discrete approach here because it is easiest to analyze. It will also be easier, taking the discrete approach, to grasp the ideas behind the mathematics.

The first step in building a one-dimensional, discrete model of energy transfer in the atmosphere is to divide the atmosphere into appropriate discrete layers or zones separated by concentric shells. We will assume that the energy content of each zone is in a steady state. Because the model ignores latitudinal and longitudinal variation, this is a reasonable approximation.[26] Of course, if the globally averaged climate is slowly changing, the steady-state assumption will not be valid. But the first step is to build a steady-state model and see what the steady state looks like; afterwards, changes in the climate can be analyzed by tinkering with the parameters characterizing the steady state (see Problems III.7 and 8).

In the steady state, the rate of energy inflow to each zone must equal the rate of energy outflow. The zones ought to be defined to facilitate this steady-state assumption. This is achieved if each zone contains one *radiation thickness* of air, i.e. the thickness of air molecules in which a quantum of IR will, on the average, be absorbed only once, the reradiation passing through to an adjacent zone. In zones one radiation thickness deep, most of the IR absorbed by each will reradiate both upward and downward to zones above and below, where the reradiated IR will again be absorbed. The steady-state condition tells us that the energy absorbed in one zone will equal the energy reradiated from that zone. Moreover, since the reradiation will be isotropically emitted (i.e., with no preferred direction), half the reradiated energy will enter the zone above and half will enter the zone below, where the process will be repeated.

The incoming solar flux initiates this sequence of steps. In order to write a set of energy-balance equations, the part of the earth-atmosphere system where this solar flux is absorbed must be specified. Figure III-5 shows how the steady-state conditions arise for an atmosphere containing n zones, under the simplifying assumption that the nonreflected solar flux is absorbed entirely at the planetary surface. At the top of the figure, above the top zone, the incoming solar flux, $\Omega/4$, enters the earth-atmosphere system; quantities $a(\Omega/4) + \sigma T_0^4$ of radiant energy leave. Here, a refers to Earth's albedo

---

26. That is, when the northern hemisphere is cold, the southern is warm, and the nightside longitudes balance out the dayside longitudes as well.

space

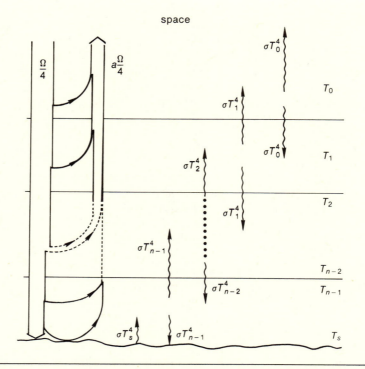

Figure III-5 The flow of energy in an *n*-layer atmosphere. The solar flux is $\Omega/4$, *a* is the albedo, and the σT^4 terms are infrared radiation fluxes (adapted from Goody and Walker, 1972).

(0.3), $a(\Omega/4)$ refers to reflected visible light, and σT_0^4 is the emitted IR radiation produced from absorbed visible light. Equating these gives

$$\Omega/4 = a\Omega/4 + \sigma T_0^4, \tag{1}$$

which is the equation determining the blackbody temperature, T_0. You'll recognize this as Eq. 4, derived in Problem II.13.

Further quantitative information is derived by requiring that energy is conserved in the topmost layer of the atmosphere. The IR flux absorbed in the top zone is σT_1^4, which is the radiation beamed upward from the zone just below. The IR flux emitted from the zone is $2\sigma T_0^4$, half of which is emitted upward and half downward.[27] With our assumption that all the nonreflected solar flux is absorbed at Earth's surface, no solar flux is absorbed in this zone, so the steady-state condition is

$$2\sigma T_0^4 = \sigma T_1^4. \tag{2}$$

---

27. The factor of 2 arises because every *surface* of a blackbody, at temperature T, radiates a flux σT^4; the zones have two surfaces, a top and a bottom.

In the second zone from the top, energy conservation tells us

$$2\sigma T_1^4 = \sigma T_0^4 + \sigma T_2^4. \tag{3}$$

Relations of this type will hold for each zone, down to the next to last. For the bottom zone, with temperature T_{n-1}, the steady-state condition is

$$2\sigma T_{n-1}^4 = \sigma T_{n-2}^4 + \sigma T_s^4, \tag{4}$$

where T_s is the surface temperature. Up to this point, there are $n + 1$ equations for $T_0, T_1, \ldots , T_{n-1}$ and T_s; thus all the atmospheric zonal temperatures and the surface temperature are determined. A further equation,

$$\sigma T_s^4 = \Omega/4 (1 - a) + \sigma T_{n-1}^4 \tag{5}$$

can be derived by conserving energy across the interface between surface and atmosphere, but it provides no new information (it can be derived from the other $n + 1$ equations).

The equations are easily solved. Eq. 1 gives

$$T_0 = [\Omega(1 - a)/4\sigma]^{1/4}. \tag{6}$$

Eq. 2 gives

$$T_1 = 2^{1/4} T_0. \tag{7}$$

Eq. 3 gives

$$T_2 = 3^{1/4} T_0, \tag{8}$$

and so on down to

$$T_s = (n + 1)^{1/4} T_0. \tag{9}$$

How large is n for Earth's atmosphere? Three kinds of molecules are responsible for most of the atmospheric absorption of IR. In order of their importance they are water vapor, carbon dioxide, and ozone. Other gases, such as methane, also absorb IR but are less important in Earth's present atmosphere. From the measured atmospheric concentrations of IR absorbing substances and from data on molecular cross sections (the latter provide a quantitative measure of the effectiveness with which individual molecules absorb IR), it has been determined that Earth's atmosphere is about two radiation zones thick.[28] Thus $n = 2$. Substituting $n = 2$ into Eq. 9 gives

$$T_s = 3^{1/4} T_0 = 3^{1/4} (255 \text{ K}) = 336 \text{ K}, \tag{10}$$

and from Eq. 7,

$$T_1 = 2^{1/4} T_0 = 303 \text{ K}. \tag{11}$$

The surface temperature of Earth[29] is actually about 290 K and not 336 K. Our one-dimensional model has greatly exaggerated the surface-to–upper atmosphere temperature difference. Eq. 9 would require an n of only 0.7 to give the correct value for T_s. A value of 0.7 is well below the measured value for n.

It is not surprising that the model[30] just presented gave an incorrect surface temperature for Earth, since several important facts were ignored. The major defects of the model are:

1. Some of the solar flux is absorbed in the atmosphere rather than at Earth's surface.

2. Some IR emitted upward from Earth's surface is not absorbed in the atmosphere; it goes directly out to space. This is mainly the IR in the wavelength band from 8 to 12 microns. The atmosphere is said to have a "window" in this range of wavelengths, being nearly transparent to IR there. Moreover, the atmosphere does not emit much IR in this wavelength interval.[31]

3. Some heat is transferred from the surface to the atmosphere by upward convection and by latent heat transfer. The model assumes radiation transfer, alone, removes heat from the surface.

These factors act to lower the surface temperature of Earth below 336 K because they lead either to less heat input to the surface or to a greater rate of heat removal from the surface.

The model is easily altered to include these corrections. To do so, we must first specify a bit more about the two atmospheric radiation

---

28.   Calculating the value of n is beyond the level of this book. However, knowing its value opens up solutions to an array of problems.
29.   Often a value of 288 K is quoted for Earth's surface temperature, but this is the near-surface air temperature. The quantity T_s characterizes the temperature of the solid and liquid surface. Because ocean surface temperature averages about 2 K above surface air temperature, $T_s = 290$ K.
30.   The model used above is discussed in a text by Goody and Walker (1972) which also contains useful information about atmospheres on other planets of the solar system. Other recommended texts which discuss planetary atmospheres and one-dimensional radiation models are Houghton (1977), Chamberlain (1978), and Walker (1977).
31.   This is an example of a general principle in physics: If an object is a good emitter in some range of wavelengths, it will also be a good absorber in that range. Conversely, poor emitters are poor absorbers.

zones on Earth. Most of the IR-absorbing capacity of the atmosphere is due to water vapor. (A way of seeing why this is so is discussed in Problem III.8.) Therefore the boundary between the lower zone and the upper zone should be very roughly at the altitude where half the atmospheric water vapor lies below and half lies above—at about 1.7 km (see Exercise 6 below). The actual boundary between zones is higher, at about 1.8 km, because IR-absorbing gases such as CO_2 are distributed in the atmosphere in such a way that their average density is found at higher elevation than water vapor's. At an altitude of 1.8 km, atmospheric pressure is about 80% of sea-level pressure, which means that about 80% of the atmosphere, by weight, lies above that altitude and about 20% lies below (see Exercise 5 below). With this information in hand, the model can be corrected. The three defects are taken up, in turn.

1. About 86 W/m^2 of solar energy (as opposed to IR) are absorbed in the atmosphere. Some of this is absorbed from the incoming beam of sunlight and some from the sunlight reflected off Earth's surface. According to the Appendix (VII.1), about 80% of the 86 W/m^2 is absorbed by dry air and 20% by water in the atmosphere, largely in cloud form. Because roughly 80% of the air and 50% of the water vapor are in the upper atmospheric zone, the fraction of the 86 W/m^2 absorbed in that zone is about $(0.80 \times 0.80) + (0.5 \times 0.20) = 74\%$. About 26% is then absorbed in the lower zone. This estimate exaggerates the difference between absorption of sunlight in the two layers because dust, which is also an absorber, lies lower in the atmosphere than does air. So we will round off the 74% to 70% and the 26% to 30%. The equations of energy conservation are easily modified to include this, but we will wait until we've discussed all the corrections before writing the revised model's equations.

2. About 20 W/m^2 of IR emitted from the surface is radiated directly to outer space through the IR windows in the atmosphere. This can be represented, very approximately, by simply adding 20 W/m^2 to the IR flux to space and subtracting it from the IR flux absorbed in the lowest IR-absorbing zone of the atmosphere. (Exercise 6 suggests a more realistic approach to including this effect in our climate model.)

3. Heat flows from Earth's surface to the atmosphere via convection and latent heat transfer. Convective heat flow is about 17 W/m^2. If the mean surface temperature were 337 K rather than 290 K, this convective heat flow would be even greater. The lapse rate of the troposphere, defined as the negative of the rate of change of temperature with increasing altitude, is about 6.5 K/km. The model that led to Eqs. 10 and 11 predicts a lapse rate of about 30 K/km. (To see this, you have to remember that T_1 is the average temperature of the lower zone, which is roughly characterized by the temperature midway into the zone at about 1.25 km altitude.) Such a

large lapse rate would result in convective upwelling of heat at a rate far exceeding 17 W/m². This upwelling, in turn, reduces the lapse rate until an equilibrium is reached at about 17 W/m². The second mechanism, latent heat released to Earth's atmosphere, occurs at a rate easily derived from two pieces of information found in the Appendix (VI.3, 1): The average annual global precipitation rate, which must equal the average annual global evapotranspiration rate; and the amount of heat needed to vaporize water that originates at 290 K. The result (see Exercise 1) is 80 W/m². Because this heat flux results in the deposition of water vapor in the atmosphere, and because the atmospheric stock of water vapor is distributed roughly equally between the lower and upper zones, it is sensible to assume that this flux is loaded into the two zones in equal measure (40 W/m² in each zone). The convective heat flux is assumed to be loaded entirely into the lower zone, although the term is small enough that this assumption makes little difference to the value of T_s.

Figure III-6 illustrates all of these energy flows in the atmosphere. The revised model can now be written. For the sake of completeness, however, we will add one more correction: a flux of waste heat, W, produced at the surface of Earth (for example, by the consumption of fossil or nuclear fuels) and loaded into the lower zone. Rewriting Eqs. 1, 2, and 4, the revised equations are:

$$W + \Omega/4 = a(\Omega/4) + \sigma T_0^4 + F_w, \tag{12}$$

$$2\sigma T_0^4 = \sigma T_1^4 + 0.5F_e + 0.7F_s, \tag{13}$$

and

$$2\sigma T_1^4 = \sigma T_0^4 + \sigma T_s^4 - F_w + F_c + 0.5F_e + 0.3F_s + W. \tag{14}$$

In these equations, F_w is the portion of IR emitted from the surface that is radiated directly to space, 20 W/m²; F_s is the portion of the solar flux absorbed in the atmosphere, 86 W/m²; F_e is the flux of latent heat leaving Earth's surface, 80 W/m²; F_c is the flux of convective heat leaving Earth's surface, 17 W/m²; a is Earth's albedo, or 0.3; and $\Omega/4$ equals 343 W/m².

Ignoring W for now (but see Problem III.9), we obtain the following solutions to these equations:

$$\sigma T_0^4 = 220.1 \text{ W/m}^2; \ T_0 = 249.6 \text{ K}, \tag{19}$$

$$\sigma T_1^4 = 340 \text{ W/m}^2; \ T_1 = 278.3 \text{ K}, \tag{20}$$

and

$$\sigma T_s^4 = 397.1 \text{ W/m}^2; \ T_s = 289.3 \text{ K}. \tag{21}$$

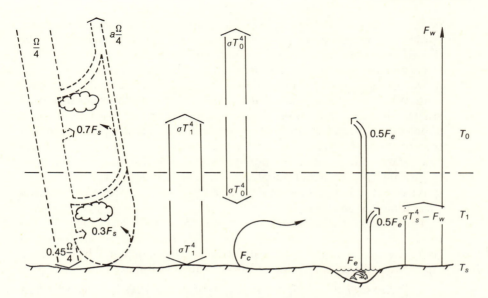

Figure III-6 Energy flows among Earth's surface, atmosphere, and space. The horizontal dashed line separates the two zones of the atmosphere (see text). T_0 and T_1 are the temperatures of these zones, and T_s is the surface temperature. The top of the diagram represents outer space. $\Omega/4$ is the incoming solar radiation (343 W/m^2), a is the albedo (≈ 0.3), and F_s is the portion of the solar flux absorbed in the atmosphere (≈ 86 W/m^2). The $0.7\,F_s$ and the $0.3\,F_s$ terms include all the absorption, from the beam of reflected solar flux as well as from the incoming beam; multiple scattering is not shown explicitly in the diagram. The solid arrows represent heat flows. σT^4_0, σT^4_1, and σT^4_s are the IR fluxes in and out of the zones; F_c is the convective heat transfer from the surface to the atmosphere (≈ 17 W/m^2); F_e is the flow of latent heat from evaporating surface water (≈ 80 W/m^2); and F_w is the IR flux that goes directly from surface to space (≈ 20 W/m^2). T_0, T_1, and T_s are determined by energy conservation (Eqs. 12-14).

These values are much more realistic than those in Eqs. 10 and 11. Note that the lapse rate, which can be estimated by recalling that the boundary between the two atmospheric zones is at an altitude of about 1.8 km, is in the right ball park. Most of the input data (the F's and a) used to calculate these values are uncertain to $\pm 10\%$, n is only approximately 2, and the atmosphere is not a perfect IR blackbody, so for this level of model-building, we have not done badly.

EXERCISE 1: Starting with an estimate of the amount of global annual precipitation, show that the latent heat flux from Earth's surface to the atmosphere is about 80 W/m^2.

EXERCISE 2: The observed tropospheric lapse rate ($-dT/dz$, where T is air temperature and z is altitude) of 6.5 K/km is nearly equal to a quantity defined as the adiabatic lapse rate. If a parcel of air near the surface rises, it will find itself surrounded by air at lower atmospheric pressure and will expand until its pressure matches that of its surroundings. In expanding, it will cool. If the expansion is sufficiently rapid, there will be no time for heat to flow in or out of the parcel; the cooling will result from the expenditure of internal energy in the parcel as it expands against its new environment. Such a process, in which no heat is exchanged, is called an adiabatic process. When the parcel has expanded so that it is at the pressure of its surroundings, it may have cooled to a temperature that is less than, equal to, or greater than the surrounding air. If its temperature *equals* that of the surrounding air, we say that the atmosphere has an adiabatic lapse rate. It can be shown (see Exercise 10) that the adiabatic lapse rate for dry air is equal to g/c_p, where g is the acceleration of gravity and c_p is the specific heat of air at constant pressure. What is the numerical value of g/c_p?

The observed average lapse rate of 6.5 K/km is less than g/c_p because real air is humid. On the basis of the fact that as air rises and cools the water vapor in it can condense, explain why the adiabatic lapse rate of humid air will be lower than that for dry air.

EXERCISE 3:   Suppose that a layer of particles has been ejected into the stratosphere such that 60% of the incident sunlight is reflected back to space and 40% is absorbed by the layer. Considering only radiation transfer and ignoring latent and convective heat flow and the IR window, calculate T_s, T_1, T_0, and the temperature of the dust layer if the dust layer is exactly one IR radiation thickness deep and the composition of the troposphere is unchanged (i.e., use a two-layer model of the atmosphere below the high dust layer).

EXERCISE 4:   The "nuclear winter" phenomenon arises because of the large quantities of soot that would get lofted into the upper troposphere in the aftermath of nuclear war. The soot, resulting from widespread fires caused by detonating nuclear bombs, would block the solar flux and cause a darkening and freezing at Earth's surface. Absorption of sunlight by the soot would warm the upper atmosphere. To explore this phenomenon (or that of a major volcanic eruption or asteroid impact causing alteration of sunlight at Earth's surface) in our model, make the following assumptions:

1. Of the incident nonreflected solar flux, 95% is absorbed in the upper zone of the atmosphere (because that is where the soot is).

2. Of the remaining 5%, half is absorbed in the lower zone of the atmosphere and half at Earth's surface.

3. Earth's albedo is unchanged, but $F_w = 0$.

4. F_e will be reduced by 50% (because the cooler surface will result in less evaporation) and F_c will equal zero (because the driving force behind vertical convection—a surface that is warmer than the air aloft—will be eliminated).

5. The soot is a poor absorber of IR, so n is still equal to 2.

What are T_s, T_1, and T_0 in steady state under these conditions? In reality, a steady-state nuclear winter would not set in because soot would gradually settle out of the atmosphere, after which the surface of Earth would warm up. This exercise also ignores the fact that reflective dust would be lofted into the stratosphere as a result of the detonations. The actual percentage of sunlight absorbed by the soot and reflected by the dust would depend on the number, size, and target sites of the detonated bombs. Should nuclear war erupt today, given the present nuclear arsenals of the United States and the Soviet Union, the value for sunlight absorption used here is not at all implausible.[32]

   * *EXERCISE 5:*   The equation of state relating pressure, P, density, ρ, and temperature, T, of an ideal gas of molecular weight, M, is $MP = \rho RT$. R is the ideal gas constant (see Appendix II).

   (*a*) Using this equation, and assuming a static atmosphere (that is, one in which the gravitational force tending to pull a patch of air downward is balanced by the upward force resulting from the pressure differential between the top and bottom of the patch), derive the following relation between pressure and altitude, z:

$$P(z) = P(0)e^{-\int_0^z dz'/H(z')},$$

where $H = RT/Mg$ and g is the acceleration of gravity.

   (*b*) If $T(z)$ in the troposphere is given by its approximate value $T(z) = 288\text{ K} - bz$, with $b = 6.5\text{ K/km}$, determine the functional dependence of P on z in the troposphere.

   (*c*) About what percentage of the mass of the atmosphere lies in the troposphere, assuming the average height of the troposphere is 12 km? What percentage of the mass of the atmosphere lies below 1.8 km? At what altitude does half the mass of the atmosphere lie below? (Hint: Remember that the atmospheric pressure at an altitude z is $M(z)g/A$, where $M(z)$ is the mass of all the air lying above altitude z, g is the acceleration of gravity, and A is Earth's area at altitude z.

   (*d*) When you answered the questions in (*c*) you probably ignored the fact that the continents displace some of the atmosphere because the continental surface is not at sea level. If the average height of the continents is 0.8 km, what are the corrected answers?

   **EXERCISE 6:*   An empirical formula for the relative humidity, h_R, of the troposphere is $h_R(z) = 0.77\ [P(z) - 0.02]/[P(0) - 0.02]$, where P is in units of atmospheres. Using that relation, the lapse rate of $(6.5\text{ K/km})z$, the tabulated relation between absolute and relative

---

32.   See Turco et al (1983) and Ehrlich et al (1983) for a more detailed discussion of the nuclear winter phenomenon and its ecological consequences.

humidity as a function of temperature (which can be found in physics and chemistry handbooks), and the fact that the atmosphere contains about 1.3×10^{16} kg of H_2O, determine the altitude below which half the water vapor lies.

EXERCISE 7: The fact that some IR emitted from the surface escapes through the 8–12 micron window of the atmosphere can be represented, in an approximate manner, by introducing a grayness factor, ϵ, corresponding to the fact that the atmosphere is not a perfect blackbody. Only a fraction, ϵ, of the IR emitted from the surface is absorbed in the lower zone of the atmosphere. Moreover, the emission of IR from the zones of the atmosphere is now a factor of ϵ times the blackbody value. The specific modifications of the energy conservation equations to include the grayness factor can be determined from the following considerations. Emission from the surface, σT_s^4, is unmodified because there is no window in the surface. In other words, the surface is still assumed to be a perfect blackbody in the infrared portion of the spectrum. However, of the IR emitted from the surface, only a fraction, ϵ, is absorbed in the lowest IR-absorbing zone of the atmosphere. A fraction, $1-\epsilon$, goes through the window (in both atmospheric layers) and is emitted directly to outer space. IR is beamed downward from the lower layer to the surface at a rate, $\epsilon\sigma T_1^4$, and all of that radiation is absorbed. Although the lower layer is transparent to radiation in the 8–12 micron window, the surface receives no IR from the upper layer because the upper layer does not emit in the range of wavelengths where the lower layer is transparent. The rates of IR emission from the upper and lower layers are now $2\epsilon\sigma T_0^4$ and $2\epsilon\sigma T_1^4$, respectively. The fraction of IR emitted from one layer toward the other and absorbed in the other is unity (again because none of the IR emitted from the atmosphere is in the 8–12 micron interval). Consideration of the case $\epsilon = 0$ will convince you that the terms F_s, F_c, and F_e in Eqs. 13 and 14 should appear multiplied by ϵ now.

Write the energy conservation equations that describe this model. Show that $\epsilon \cong 0.9$ corresponds to $F_w = 20$ W/m^2. What values of T_0, T_1, and T_s does this "gray body" model yield if $\epsilon = 0.9$?

** *EXERCISE 8:* Using the basic approach developed in Problem III.5, construct a more complex box model to describe the latitudinal temperature variation on Earth's surface and in the two atmospheric zones. You might divide the northern and southern hemispheres each into two latitudinal zones (0°–45°; 45°–90°). The following considerations are relevant:

1. Incident solar flux will be less at the poleward zones than at the low latitude zones. A trigonometric factor describes this.

2. Both seawater and atmosphere convect heat from the equator toward the poles (see Appendix XIV.1).

3. There is considerably more IR-absorbing water vapor (and therefore n will be greater) in the low-latitude atmospheric zones than in the poleward atmospheric zones. As a rough guess, try $n = 3$ between 0° and 45°, $n = 1$ between 45° and 90°. Better yet, consult Houghton (1977) to determine how much atmospheric water vapor is above and below the 45th parallel. Very ambitious readers may want to estimate how much water is above and below the 45th parallel by using the empirical formula in Exercise 6 and solving self-consistently for the temperature in the two low- and high-latitude zones.

** *EXERCISE 9:* Eqs. 1–5 are the steady-state conditions that would arise if a dynamic model for the time dependence of $T_0, \ldots,$ T_{n-1} and T_s were constructed and then evaluated at $dT_0/dt = 0, \ldots,$ $dT_{n-1}/dt = 0$, and $dT_s/dt = 0$. Construct such a dynamic model for the case $n = 2$ and use it, along with the method of perturbation analysis presented in Problem III.5, to calculate how T_s, T_0, and T_1 would behave, in time, during the first year following a permanent 10% reduction in the solar flux. You will understand all the essentials if you work with a dynamic model whose steady-state limit is Eqs. 1, 2, and 5 rather than the more complex Eqs. 12–14. Each differential equation will be of the form $K_i dT_i/dt = Q_{i,in} - Q_{i,out}$, where $i = 0$, 1, or s (for surface). $Q_{i,in}$ is the rate of energy inflow, per unit area, to the ith zone; $Q_{i,out}$ is the rate of energy outflow, per unit area, from the ith zone; and K_i is the heat capacity, per unit area, of the ith zone.

To set up the differential equations, you will have to estimate the heat capacities, K_i, of the two atmospheric zones and of Earth's surface. The heat capacities of the atmospheric zones can be estimated knowing the heat capacity of air at sea-level pressure and your result from Exercise 5c. The heat capacity of Earth's surface depends on what depth of land and water one considers. That choice, in turn, depends on what time scales of change one is interested in. The heat capacity of the mixed layer of the oceans governs the changes that occur over a period of months to decades, while soil and rock govern changes over periods of days to months. Much more rapid temperature changes (on the order of days or minutes) are described by the heat capacity of buildings and bodies. To describe long-term effects, on the order of centuries, the heat capacity of the deep ocean must be used. For simplicity, in your model assume that the entire Earth is covered by ocean. You will have to estimate the heat capacities of the oceanic and atmospheric layers.

** *EXERCISE 10:* Derive the result that the adiabatic lapse rate for dry air (see Exercise 2) is $dT/dz = -g/c_p$.

7. Land Use and Climate

Suppose that 20% of the land area of Earth is deforested and the area subsequently desertifies. By about how much would Earth's average surface temperature change?

· · · · · · ·

Desert has a higher albedo than forest. Therefore, deforestation and subsequent desertification will increase Earth's albedo, which in turn will cool Earth's surface. If a value of Earth's albedo, a, different from 0.3 is substituted into Eqs. 12–14 of Problem III.6, revised values for T_s, T_0, and T_1 can be calculated directly. Deforestation can alter climate in other ways besides changing Earth's albedo. Exercises 3 and 4 are concerned with some of these other mechanisms.

The first step is to determine the change, Δa, in Earth's albedo resulting from a change in land use. This task is not as straightforward as it might appear, however, because a given change in the albedo of the surface of Earth will generally not equal the change in the total albedo of Earth. The reason for this hinges on the definition of albedo: the ratio of reflected to incident radiation. Earth's albedo, a, is

$$a = \frac{\text{solar flux reflected from Earth to space}}{\text{solar flux incident on Earth}}, \tag{1}$$

while the albedo, R_S, of Earth's surface is

$$R_S = \frac{\substack{\text{solar flux reflected from Earth's surface} \\ \text{to the atmosphere}}}{\text{solar flux incident on Earth's surface}}. \tag{2}$$

Note that a will only be influenced by R_S to the extent that light reflected off the ground penetrates the atmosphere and passes through to space. In the limiting case where no incident sunlight penetrated the atmosphere, a given change in R_S would lead to no change at all in a because the total albedo of the Earth in that special case would equal only the atmospheric albedo.

Some fraction of the light reflected up from Earth's surface will not escape directly to space. Instead it will either be absorbed in the atmosphere or reflected back to the ground. Although our climate model, as described by Eqs. 12–14 of Problem III.6, does not describe this phenomenon explicitly, neither are the two contradictory. In those equations, the albedo, a, is the total albedo of Earth and includes all the multiple scattering corrections. Similarly, F_S is the total amount of solar energy absorbed in the atmosphere, including a small portion that first bounced off the surface before being absorbed. For our purposes now, it is necessary to describe these multiple scattering effects explicitly. Otherwise, we will overestimate by a factor of about two the decrease in T_s.

To include these corrections, we use the multiple scattering formalism presented in Problem II.18 and assume that the absorption and reflection coefficients of the atmosphere are the same whether viewed from the top looking down or the bottom looking up. We let R_S and R_A be the reflection coefficients (albedos) of the surface and atmosphere, while A_S and A_A are the absorption coefficients. Using Eq. 6 of Problem II.18, and recalling that the transmission coefficient of the atmosphere, T_A, can be written $T_A = 1 - R_A - A_A$, the albedo of Earth as a whole can be written

$$a = R_A + \frac{(1 - R_A - A_A)^2 R_S}{1 - R_A R_S}. \tag{3}$$

From your result in Exercise 3 of Problem II.18, you know that f, the total fraction of the incoming flux that is absorbed in the atmosphere, is given by

$$f = A_A \left[1 + \frac{(1 - R_A - A_A)R_S}{1 - R_A R_S} \right]. \tag{4}$$

The quantities a and f are known: $a = 0.3$ and $f = 86/343 = 0.25$ (see Problem III.6). In Eqs. 3 and 4 there are 3 unknowns (R_A, A_A, and R_S), so one more piece of information is needed. Direct measurements of the transmission of the incoming beam of sunlight reveal that $A_A \approx 0.23$. Therefore, from Eqs. 3 and 4, $R_S \approx 0.16$ and $R_A \approx 0.26$.

Substituting the values of R_A and A_A (which we assume are unchanged by the deforestation) back into Eq. 3,

$$a = 0.26 + \frac{(0.49)^2 R_S}{1 - 0.26 R_S}$$
$$= 0.26 + \frac{0.24 R_S}{1 - 0.26 R_S} \tag{5}$$

With this formula, we can calculate the change in a induced by a change in R_S. If the change in R_S is sufficiently small, then $\Delta a/\Delta R_S \approx da/dR_S$. Assuming this to be the case, and differentiating Eq. 5, we get

$$\frac{da}{dR_S} = \frac{0.24}{(1 - 0.26 R_S)^2}. \tag{6}$$

Evaluated at the original, unperturbed value for R_S of 0.16, this becomes

$$\frac{da}{dR_S} = 0.26. \tag{7}$$

Hence,

$$\Delta a \approx 0.26 \Delta R_S. \tag{8}$$

What is the change in R_S resulting from the hypothesized land use changes? The Appendix (XIV.2) tells us that typical forested land has an albedo of roughly 0.15, while desert has an albedo of roughly 0.25. Since 29% of Earth's surface area is land, a change from forests to desert on 20% of the Earth's land area is equivalent to a similar change on 0.20×0.29 or about 6% of Earth's total surface area. A change of surface albedo from 0.15 to 0.25 on this 6% of Earth's surface is equivalent to a change in R_S of $0.06(0.25 - 0.15)$ or 0.006. This can be stated as

$$\Delta R_S = 0.006. \tag{9}$$

Substituting Eq. 9 into Eq. 8, we obtain

$$\Delta a = (0.26)(0.006) = 0.00156. \tag{10}$$

The new albedo resulting from deforestation is $0.3 + 0.00156 = 0.30156$. We can substitute this value into Eqs. 12–14 of Problem III.6 and calculate the altered temperatures. The results are

$$T_0 = 249.5 \text{ K}, \tag{11}$$

$$T_1 = 278.1 \text{ K}, \tag{12}$$

and

$$T_s = 289.0 \text{ K}. \tag{13}$$

Comparing Eq. 13 here with Eq. 21 in Problem III.6, we see that the average surface temperature drops by 0.3 K. Readers interested in further discussion of this topic, including estimates of the likely climatic consequences of historic albedo alteration, should look at Sagan et al. (1979).

EXERCISE 1: Verify that the approximation, $\Delta a / \Delta R_S \approx da/dR_S$, was a good one.

EXERCISE 2:   What possible mechanisms might nullify our assumption that R_A and A_A are unaltered by deforestation?

* *EXERCISE 3:*   A consequence of deforestation that was not included in the treatment above is a decrease in the rate of evapotranspiration and therefore of latent heat transfer to the atmosphere. Estimate the effect of this, alone, on T_s, T_1, and T_o, using the climate model of Problem III.6. Assume that prior to deforestation the rate of evapotranspiration, per unit area of land in question, was equal to the average land evapotranspiration rate and that it is reduced, after deforestation, to one fourth its former value. You will find that the surface warming caused by reduced evapotranspiration is roughly of the same magnitude as the cooling effect of increased albedo. Should you, therefore, not worry about the overall climate consequences of deforestation? In answering this, give some thought to the difference between global and regional climate alteration.

EXERCISE 4:   Climate is affected by the CO_2 content of the atmosphere (see Problem III.8). By roughly what percentage will the atmospheric concentration of CO_2 immediately increase if the deforestation and subsequent burning of the cleared vegetation occur so rapidly that ocean uptake of CO_2 can be ignored? See Appendix, Section XII.2; assume that only tropical and temperate forests are cut and that for every 3 km^2 of tropical deforestation, there is 1 km^2 of temperate deforestation.

* *EXERCISE 5:*   Verify the claim that, if multiple-scattering corrections are ignored, the value of Δa and ΔT_S are overestimated by about a factor of two. Give a qualitative argument why multiple-scattering corrections reduce the magnitude of the effect.

8. The CO_2 "Greenhouse" Effect

If the burning of fossil fuels were to cause the carbon dioxide concentration in Earth's atmosphere to become twice what it was at the beginning of the industrial revolution, how would Earth's surface temperature be affected?

. . . .

It is difficult to make even a wild guess at the answer to the question above, but some insights can be obtained from a striking observation: There is presently about as much carbon in the form of CO_2 in the atmosphere [735×10^{12} kg(C)] as there is in the living biosphere [about 560×10^{12} kg(C)]. Over the past few hundred million years of Earth's history, changes in the amount of carbon locked up in living biomass have undoubtedly occurred, on a scale comparable in order of magnitude to the amount present in that form today. These biospheric changes would have led to large changes in the atmospheric CO_2 concentration. Over that same period, Earth's temperature has varied as well. Climatologists believe that the range of the temperature variation was at most ± 10 K (Bolin 1974). Therefore, we do not expect an order-of-magnitude larger change (e.g., $\pm$ 100 K) in Earth's surface temperature, T_s, to result from a doubling of atmospheric CO_2. Of course, the historic changes in Earth's surface temperature may have had little to do with changes in atmospheric CO_2; in that case we would argue for an even smaller response to a CO_2 doubling.

Further evidence for a change in T_s no greater than 10 K comes from recent history. From the beginning of the industrial revolution to the present (1983), the atmospheric CO_2 concentration has increased from about 275 ppm(v) to 340 ppm(v), an increase of 24%. The average global surface temperature has increased during that period by at most 1 or 2 K. Without implying a causal connection, we can use the information to argue, again only suggestively, that an increase in temperature on the order of, at most, 10 K should result from a CO_2 doubling.

One flaw in these arguments is that the effects of CO_2 changes may have been canceled by the effects of other changing parameters in the past, while they might not be in the future. But enough wild guessing. Let's see how to calculate an approximate answer.

The direct effect of an increase in atmospheric CO_2 is to increase the amount of IR-absorbing material in the atmosphere. In the language of the climate model we developed in Problem III.6, this shows up as an increase in n—the number of IR-absorbing "boxes" in the atmosphere. If, in our climate model, expressed by Eqs. 12–14 in Problem III.6, all parameters except n are held fixed (i.e., a, W, Ω, σ, F_w, F_s, F_e, and F_c), the effect of an increase in n on T_s, and on T_1 as well, can be calculated. We will perform that calculation first. However, the result won't tell the whole story, because the other parameters in Eqs. 12–14 may change as a result of the initial change in T_s. Even n itself may change further in response to the initial increase in T_s caused by the initial increase of n. Such secondary changes are called feedback effects, and we must estimate them as well.

Our plan of attack is as follows. First we will derive a general formula relating a change in n to a direct change in T_s. Next we'll estimate the direct effect on n of a CO_2 doubling and substitute that numerical value into the general formula, thereby calculating a numerical value for the initial change in T_s. Finally we will examine the feedback effects.

It may seem peculiar to talk about a change in n. After all, n is the number of boxes into which the atmosphere is divided. Suppose the change in n due to a CO_2 doubling is, say, an increase by 10% from 2.0 to 2.2. How can we have a box model with 2.2 boxes? The answer lies in our ability to interpolate. Eq. 9 of Problem III.6 is an example of a relation between T_s and n derived for integer values of n. However, the formula, once derived, can be used for any value of n, including non-integer values.

But Eq. 9 is not good enough for us now because it does not include all the corrections used in the more complete model, expressed as Eqs. 12–14 of Problem III.6. So, consider Eqs. 12–14 with W set equal to zero (because we are not interested, here, in anthropogenic waste heat). Rewriting these equations so as to eliminate T_1 and T_0, we obtain

$$\sigma T_s^4 = 3\frac{\Omega}{4}(1 - a) - (F_c + 1.5F_e + 1.7F_s + 2F_w). \quad (1)$$

Where did n go? Referring to the derivation of Eq. 9 in Problem III.6, you can show that the factor of 3 multiplying $\Omega(1 - a)/4$ is actually $n + 1$, and the factor of 2 multiplying F_w is actually n. So let's rewrite this as

$$\sigma T_s^4 = (n + 1)\frac{\Omega}{4}(1 - a) - (F_c + 1.5F_e + 1.7F_s + nF_w). \quad (2)$$

Using the numerical values for the F's given in Problem III.6, we calculate

$$\sigma T_s^4 = (n + 1)\frac{\Omega}{4}(1 - a) - 283.2 - n(20), \quad (3)$$

in units of W/m^2.

The effect, ΔT_s, on T_s of a small change, Δn, in n can now be derived. We first write

$$T_s = \left[\frac{(n + 1)\frac{\Omega}{4}(1 - a) - 283.2 - 20n}{\sigma} \right]^{1/4}. \tag{4}$$

Then, differentiating, we obtain

$$\frac{dT_s}{dn} = \frac{\frac{\Omega}{4\sigma}(1 - a) - \frac{20}{\sigma}}{4\left[\frac{(n + 1)\frac{\Omega}{4}(1 - a) - 283.2 - 20n}{\sigma} \right]^{3/4}}. \tag{5}$$

With a little algebraic manipulation and with the use of Eq. 4, this can be rewritten as

$$\frac{dT_s}{dn} = \frac{T_s}{4(n + 1)}\left(1 + \frac{263.2}{\sigma T_s^4} \right). \tag{6}$$

If Δn and Δt_s are small compared to n and T_s, respectively, then this can be written as

$$\frac{\Delta T_s}{T_s} = \frac{\Delta n}{4(n + 1)}\left(1 + \frac{263.2}{\sigma T_s^4} \right). \tag{7}$$

With the numerical value of σT_s^4 substituted in ($\sigma T_s^4 = 397.1\ W/m^2$), we find

$$\frac{\Delta T_s}{T_s} = 0.42\ \frac{\Delta n}{n + 1}. \tag{8}$$

Next, we need to determine the value of Δn resulting directly from a doubling of the atmospheric CO_2 concentration. A realistic calculation of Δn would involve an excursion into molecular physics;[33] here, we will derive a "spherical cow" approximation.

Note, first, that a doubling of CO_2 will not double n because the value of n is determined by many atmospheric constituents, not just CO_2. Water vapor, CO_2, and ozone are the major contributors to the IR absorptivity of the atmosphere, with H_2O the most important and

---

33. Many of the nuances of the problem are discussed clearly in Ackerman (1979), Augustsson and Ramanathan (1977), Schneider (1972), and Manabe and Wetherald (1967).

CO_2 second. Our approach will be to calculate the relative fractional contribution of CO_2 to the present value of n. If that fraction is multiplied by n, the result will be the desired value of Δn resulting from a CO_2 doubling.

Don't confuse this calculation of Δn with a calculation of the change in the IR absorption coefficient (see Problem II.18 for a discussion of absorption coefficients). If the concentration of *all* the IR absorbers in the atmosphere were to double, n would roughly double; whereas the IR absorption coefficient for the whole atmosphere, which is now close to 1, would only increase slightly.

But n would not exactly double, either, because as the concentration of IR-absorbing gases increases, the effectiveness of individual molecules in absorbing IR, changes. As the gas molecules rub shoulders, and the pressure increases, the range of wavelengths within which the molecules effectively absorb, broadens. Moreover, when gases absorb IR, they can subsequently emit IR at wavelengths longer than those absorbed. In portions of the electromagnetic spectrum, CO_2 and H_2O are particularly effective at absorbing IR; the portions are different for each gas. Thus one gas may emit IR in a portion of the spectrum where another is ineffective as an absorber, introducing nonlinear effects. Indeed, an increase in certain atmospheric trace constituents, such as methane or ozone, can have a particularly large effect on surface temperature.[34] The reason is that such gases often absorb IR in the window through which H_2O and CO_2 transmit IR. Thus these gases in effect decrease the value of F_w as well as increase the value of n. For all these and other reasons, determining the effect of an increase in gas concentration on n is difficult. Here we'll attempt only a crude approximation, ignoring the complications. Exercise 7 points the way toward an estimate of ΔT_s resulting from the partial closure of an IR window.

Returning to our task of estimating Δn, the concentration of CO_2 in Earth's atmosphere prior to the beginning of the industrial revolution and the era of fossil fuel burning was about 275 ppm(v). If we multiplied this by the number of moles of air, 1.8×10^{20}, we get 49.5×10^{15} moles(CO_2). The mass of water vapor in the atmosphere is 1.3×10^{19}g or 722×10^{15} moles(H_2O). On a mole-per-mole basis, the ratio, r, of CO_2 to H_2O plus CO_2 is

$$r = \frac{49.5}{49.5 + 722} = 0.064 \qquad (9)$$

A mole of any substance contains Avogadro's number of molecules, and therefore r is a ratio of molecules as well as of moles.

---

34. Kerr (1983b) reviews recent studies on the role of trace gases in greenhouse warming.

If CO_2 and H_2O were equally effective as IR absorbers, then the value of Δn resulting from an increase in the atmospheric concentration of CO_2 would very roughly (see discussion above) equal r times n times the percentage increase in the atmospheric CO_2 concentration divided by 100. However, an individual CO_2 molecule is only about one fourth as effective as a water vapor molecule in absorbing IR (see Exercise 4), and so r must be modified. Rather than the ratio of the number of molecules of the gases we want the ratio of the IR-absorbing effectiveness of the gases. We therefore replace each 49.5 in Eq. 9 by 49.5/4. This modified r has a value of 0.0169. Hence, if the atmospheric concentration of CO_2 doubles, we calculate that

$$\Delta n = 0.0169 \times n = 0.0338. \tag{10}$$

Substituting this result into Eq. 8 gives

$$\frac{\Delta T_s}{T_s} = \frac{0.42}{3}(0.0338) = 0.0047. \tag{11}$$

Using the value $T_s = 290$ K, we determine that

$$\Delta T_s = 1.4 \text{ K.} \tag{12}$$

Given the uncertainties in this derivation, a round-number range of values from 1 to 2 K is appropriate.

The final step is to estimate the feedback effects. There are several such feedbacks to consider. A warmer surface temperature will increase the rate of evaporation of water from Earth's surface. Assuming the residence time of water in the atmosphere does not decrease to compensate for this increased inflow, this will lead to some increase in the stock of water in the atmosphere. If this water is in vapor form, n will increase, causing a further increase in T_s. This is an example of a positive feedback. If the increase in the stock of atmospheric water causes added cloud formation, however, then the albedo of Earth could increase, and this would have a cooling effect. This is an example of a negative feedback.

Cloud formation could increase if the relative humidity of the air increases. Climatologists are uncertain about whether the relative humidity will remain constant or increase if the surface temperature increases. However, relative humidity changes little from summer to winter, when large surface-temperature changes occur. This suggests, as a first approximation, that a temperature increase resulting from a CO_2 doubling would leave the relative humidity unchanged. In this case, the positive feedback effect of water vapor on n should dominate the negative feedback effect of increased cloud cover.

A third feedback effect arises in polar regions, where a sustained warmer temperature will shrink the ice caps somewhat.[35] Because ice has a higher albedo than open seawater, the shrinkage of the ice cap will lower the albedo of the polar regions, increasing the warming.

Yet another feedback mechanism involves CO_2 dissolved in seawater. As the surface temperature, T_s, increases so will seawater temperature. A warming ocean, like a warming beer, releases CO_2 to the atmosphere. As the sea goes "flat," the air becomes even warmer and in turn warms the sea.

Treatment of the seawater-CO_2 feedback effect is left to Exercise 5. The ice-albedo feedback effect is a regional one, explored in Exercise 6 below. The water vapor feedback effects are taken up in Exercises 2 and 3.[36] A general treatment of how to calculate any feedback phenomenon is presented below.

Suppose that the direct effect on T_s of an increase in n by $(\Delta n)_1$ is to increase T_s by an amount, $(\Delta T_s)_1$, as calculated from Eq. 8. Suppose further that because of a feedback effect (not included in Eq. 8), the result of the increase in T_s by $(\Delta T_s)_1$, is to further increase n so that the total increase in n is now $(\Delta n)_2$. By using $(\Delta n)_2$ rather than $(\Delta n)_1$ in Eq. 8, we would have a more accurate estimate of ΔT_s. Using consistent notation, we would call that improved value $(\Delta T_s)_2$. If

$$(\Delta n)_2 - (\Delta n)_1 << (\Delta n)_1$$

and   (13)

$$(\Delta T_s)_2 - (\Delta T_s)_1 << (\Delta T_s)_1,$$

we say that the feedback has converged rapidly. In words, Eq. 13 means that the differences between the initial and first iterative result for Δn and for ΔT_s are small compared to the initial results. If inequalities 13 are not satisfied, we must perform further iterations. A procedure for doing this systematically will be explained below.

To begin, let's describe the first iteration systematically. Let T_s equal a function, F, of n and let n be a function of T_s:

$$T_s = F[n(T_s)].$$ (14)

Eq. 4 was an example of such a function, F, although in that equation the T_s dependence of n was not made explicit because we were ignoring feedback then. Denote the values of n and T_s prior to the CO_2 doubling by n_0 and $T_{s,0}$. Let $(\Delta n)_1$ be the initial direct effect on n_0 of a

---

35. Floating icebergs will melt to smaller dimensions, but the major effect will be to increase the rate at which large ice masses decrease in area as hunks of ice calve into the sea at an increasing rate.

36. Further discussion of the CO_2 greenhouse problem and of feedback effects stemming from a CO_2 increase may be found in Clark (1982). This reference also offers a hefty bibliography of journal articles on climate and CO_2.

CO_2 doubling, and let $(\Delta T_s)_1$ equal the initial effect on T_s of the increase in n_0 by $(\Delta n)_1$. Then we can write

$$(\Delta T_s)_1 = \left. \frac{\partial F}{\partial n} \right|_{\substack{n = n_0 \\ T_s = T_{s,0}}} (\Delta n)_1.$$
(15)

Eq. 8 is just a particular example of Eq. 15. To include the first feedback correction, we write

$$(\Delta T_s)_2 = \left. \frac{\partial F}{\partial n} \right|_{\substack{n = n_0 \\ T_s = T_{s,0}}} (\Delta n)_1 + \left. \frac{\partial F}{\partial n} \frac{\partial n}{\partial T_s} \right|_{\substack{n = n_0 \\ T_s = T_{s,0}}} (\Delta T_s)_1.$$
(16)

Eq. 16 is by no means an exact result for $(\Delta T_s)_2$. Even if there were no feedback, so that the second term on the right-hand side of Eq. 16 were zero, the remaining term on that side of the equation would still be only the first term in a Taylor series expansion[37] for ΔT_s. Similarly, the feedback term is only the first term in another Taylor series expansion. As long as $(\Delta n)_1 << n_0$ and $(\Delta T_s)_1 << T_{s,0}$ these truncated Taylor series expansions are a good approximation.

Even if $(\Delta n)1 << n_0$, and $(\Delta T_s)1 << T_{s,0}$, however, the convergence of the feedback may be slow if inequalities 13 are not satisfied. In this case, Eq. 16 can be improved in accuracy by replacing $(\Delta T_s)_1$ by the more exact result $(\Delta T_s)_2$:

$$(\Delta T_s)_2 = \left. \frac{\partial F}{\partial n} \right|_{\substack{n = n_0 \\ T_s = T_{s,0}}} (\Delta n)_1 + \left. \frac{\partial F}{\partial n} \frac{\partial n}{\partial T_s} \right|_{\substack{n = n_0 \\ T_s = T_{s,0}}} (\Delta T_s)_2.$$
(17)

This can be rewritten as

$$(\Delta T_s)_2 = \frac{\left. \dfrac{\partial F}{\partial n} \right|_{\substack{n = n_0 \\ T_s = T_{s,0}}} (\Delta n)_1}{(1 - A)},$$
(18)

---

37.   See Swartz (1973) for a review of the Taylor series expansion.

where

$$A = \frac{\partial F}{\partial n}\frac{\partial n}{\partial T_s}\bigg|_{\substack{n = n_0 \\ T_s = T_{s,0}}}.$$

(19)

The term $(1 - A)^{-1}$ is called the feedback factor. Note that if A is negative, the feedback is negative and $(\Delta T_s)_2 < (\Delta T_s)_1$.

Eq. 17 describes only one class of feedback mechanism—that mediated by a change in n. The second term on the right-hand side of that equation can be replaced by a more general term,

$$\frac{\partial F}{\partial n}\frac{\partial n}{\partial T_s}\bigg|_{\substack{n = n_0 \\ T_s = T_{s,0}}} (\Delta T_s)_2 \rightarrow \sum_i \frac{\partial F}{\partial p_i}\frac{\partial p_i}{\partial T_s}\bigg|_{\substack{p_i = p_{i,0} \\ T_s = T_{s,0}}} (\Delta T_s)_2 \ ,$$

(20)

where the p_i are any climate parameters, including n, upon which the function, F, depends. Thus A, in Eq. 19, has a more general structure:

$$A = \sum_i \frac{\partial F}{\partial p_i}\frac{\partial p_i}{\partial T_s}\bigg|_{\substack{p_i = p_{i,0} \\ T_s = T_{s,0}}}.$$

(21)

One of the p_i, for example, could be the surface albedo of Earth, which will decrease if T_s increases and the ice cover shrinks (see Exercise 6). Similarly, the numerator in Eq. 18 is only a specific example of what is called a driving term. In this case, the driving force is the Δn caused directly by a CO_2 doubling. The driving force could be, for example, an albedo change due to deforestation (see Problem III.7).

Using Eqs. 18 and 19, we can estimate the feedback corrections to our result, Eq. 12. Exercises 2, 3, 5, and 6 offer an opportunity for you to do so.

   * *EXERCISE 1:*   It has been estimated[38] that a doubling of the atmospheric concentration of CO_2 would appreciably reduce the rate of water transpiration by green plants. Assume that the rate of evapotranspiration from the land surfaces of Earth decreases by 5% from

---

38. See Strain and Armentano (1980) for a discussion of this effect and other consequences for the biosphere of enhanced CO_2 levels.

this effect, and ignore, for now, a possible increase in evaporation rate due to an increased T_s. If the residence time of water vapor in the atmosphere remains constant, would the combined direct effect of the CO_2 doubling on n and the indirect effect of altered transpiration on the water vapor content of the atmosphere (and therefore on n) cause n to increase or decrease?

  * EXERCISE 2:   Calculate the feedback effect on ΔT_s if the evapotranspiration rate increases by 2% for every 1 K increase in T_s. You can ignore the feedback effect on n arising from a change in atmospheric water vapor content, focusing your effort here on the fact that F_e in Eqs. 12–14 of Problem III.6 will increase, thereby tending to *cool* Earth's surface.

  ** EXERCISE 3:   Ignore the changes in evapotranspiration and estimate the magnitude of the feedback effect resulting from the change in atmospheric water vapor that is expected if the temperature of the atmosphere increases and relative humidity stays fixed. Note that the change in absolute humidity will be determined by T_1 (see Problem III.6) rather than T_s. Hence, Eqs. 12–14 of Problem III.6 will first have to be used to calculate the change in T_1 resulting directly from the change $(\Delta n)_1$ caused by the CO_2 doubling. You can then calculate the feedback effect on T_s by modifying Eq. 16, slightly, to read

$$(\Delta T_s)_2 = \left.\frac{\partial F}{\partial n}\right|_{\substack{n = n_0 \\ T_s = T_{s,0}}} (\Delta n)_1 + \left.\frac{\partial F}{\partial n}\frac{\partial n}{\partial T_1}\right|_{\substack{n = n_0 \\ T_s = T_{s,0} \\ T_1 = T_{1,0}}} (\Delta T_1)_1.$$

The value of $\partial n/\partial T_1$ will have to be estimated from the relation between absolute and relative humidity as a function of temperature (see tabulations in any handbook of chemistry or physics).

  **EXERCISE 4:   Derive the result that a CO_2 molecule is about one fourth as effective an IR absorber as is an H_2O molecule, for a blackbody spectrum characteristic of Earth's surface or atmosphere. Each gas species has a characteristic function describing the likelihood that individual molecules will absorb IR radiation; this function, called a cross section, depends on the wavelength of the radiation. For any given spectrum of wavelengths, the overall absorption effectiveness of a molecular species can be determined. The spectrum of wavelengths for a blackbody can be characterized by specifying the temperature at which the blackbody emits. For the case at hand, we can take the blackbody temperature to be 280 K—the approximate value of the blackbody temperature of the lower atmosphere. To simplify the problem, approximate the cross section for absorption by

CO_2 and H_2O by step functions that take on values of either 0 (no absorption) or 1 (complete absorption). You can assume that CO_2 absorbs completely in the wavelength range from 13 to 16 microns, and that water vapor absorbs completely in the wavelength ranges from 5 to 7 and 13 to ∞ microns. The task then is to integrate the 280 K blackbody energy spectrum over these wavelength ranges and take the ratio of the amount of energy in the CO_2-absorbing range to that in the H_2O-absorbing range. See Rohsenow and Hartnett (1973) for tabulated blackbody spectra. Crude numerical integrations over the wavelength ranges can be performed using these spectra.

Note again that this estimate results from an extreme oversimplification.

** *EXERCISE 5:* Combining the treatment of seawater-CO_2 chemistry in Problem III.4 with the treatment of the CO_2-temperature interaction used here, estimate the magnitude of the seawater-CO_2 feedback on T_s. Assume the diffusion of CO_2 between air and sea is linear, donor-controlled and that seawater alkalinity is constant. You will need to look up in a text such as Stumm and Morgan (1981) the temperature dependence of the constants K_H, K_1, and K_2 used in Problem III.4.

* *EXERCISE 6:* One consequence of a greenhouse warming is the melting of polar ice. Concern has been raised that this might result in an elevated sea level. A rise in sea level would be brought about by the melting of nonfloating ice, such as the ice shelves comprising much of Greenland and Antarctica. The melting of ice floating in the sea (sea ice) will not raise sea level but it will reduce the albedo of the polar region and thereby raise the polar surface temperature further. To estimate the magnitude of this feedback effect, we must calculate the feedback factor,

$$(1 - A)^{-1} = \left. \left(1 - \frac{\partial F}{\partial a} \frac{\partial a}{\partial T_s} \right)^{-1} \right|_{\substack{a = a_0 \\ T_s = T_{s,0}}},$$

where a is the polar albedo, T_s is the polar surface temperature, and $T_s = F[a(T)]$.

Doing this accurately is difficult because global averaging will not suffice to calculate a regional effect. So let's try a spherical cow approach. Consider a planet that all over has the same optical properties as the polar regions on Earth—namely, the reflection coefficient of the atmosphere is 0.3, the total albedo is 0.6, and the transmission coefficient of the atmosphere is 0.6. The planet's mean temperature is 222 K. Since a 50 K rise in the average polar temperature would result in the permanent melting of most of the sea ice, we will assume that the areal extent of sea ice is reduced by 2% for every °C increase in surface temperature on this planet. Using a simple $\sigma T^4 = (\Omega/4)(1 - a)$ model, calculate the feedback factor, assuming that the water beneath the ice cover has an albedo of 0.2. To evaluate da/dT you will have to use the chain rule: $da/dT = (da/dR_s)(dR_s/dE)(dE/dT)$, where R_s is the surface albedo, and E is the areal extent of ice. The value of R_s and of da/dR_s can be determined using the methods of Problem III.7. If the albedo of the water beneath the ice had been zero instead of 0.2, A would equal 1.04. How would you interpret a situation in which $A > 1$?

EXERCISE 7: An important mechanism by which atmospheric gases can affect T_s is by partially closing IR windows. This effect can be explored by estimating how a change in the grayness parameter, ϵ (see Exercise 7 of Problem III.6), affects T_s. Starting with the model derived in Exercise 7, Problem III.6, show that $\Delta T_s/T_s = 0.24\,\Delta\epsilon/\epsilon$.

9. Urban Heat Islands

How much warmer is a densely populated urban area than the surrounding countryside?

· · · · · · ·

To be specific, consider a square urban area, 20 km on a side, with a population of 10^7. The residents consume energy at the average per capita rate for the United States in 1980. As a way of simplifying the problem, consider first only the warming caused directly by the heat generated in the urban area. Later, we'll discuss the other ways a city influences its local climate, and we'll determine their qualitative influence on our answer.

A plausible, simple, and (as seen below) wrong approach to the problem is to make direct use of the results of Problems II.13 and III.6. In Problem II.13 it was shown that Earth's blackbody temperature, T_0, is related to the solar flux by

$$\sigma T_0^4 = \frac{\Omega}{4}(1 - a). \tag{1}$$

An increase in the rate of energy input to the atmosphere, W, would increase T_0 to a new value, T_0', given by

$$\sigma(T_0')^4 = \frac{\Omega}{4}(1 - a) + W. \tag{2}$$

In the same spirit, the change in the surface temperature, T_s, can be estimated with the one-dimensional climate model developed in Problem III.6. The relation that can be derived (see Exercise 1) between W and the new surface temperature, T_s', is:

$$\sigma[(T_s')^4 - T_s^4] = 2W. \tag{3}$$

How big is the anthropogenic term W? Using the U.S. population and energy consumption data from the Appendix, we can calculate that the urban area generates heat with a power density of

$$W = \frac{(11.2 \times 10^3 \text{ watts/person}) \times (10^7 \text{ people})}{(20 \text{ km})^2 (10^3 \text{ m/km})^2} \tag{4}$$
$$= 280 \text{ watts/m}^2.$$

Substitution into Eq. 2 yields $T_0' = 309$ K. Substitution into Eq. 3 yields $T_s' = 361$ K, which is 71 K hotter than Earth's present average surface temperature of 290 K!

This is an absurd result. If it were true, residents of New York City would be stewing. What has gone wrong? Actually, nothing, provided the urban area with the specified density of energy consumption covered all of Earth. For then, only upward radiation would dissipate the anthropogenic heat, and our calculation would generate a good approximation to T_0' and T_s'. But when an urban area is fairly small compared to the non-urbanized area, the major mechanism of heat removal is not upward radiation but horizontal, wind-generated, convective dissipation, preceded by rapid vertical turbulent mixing of the surface-generated heat to an altitude where the winds are effective (see Figure III-7). If the urban area covered the planet, the winds would only transfer heat from one spot to another. They could not dissipate it because there would be no cooler air for hot air to mix with.

With the mechanism of urban heat dissipation identified, the problem can now be solved. Our plan of attack is to determine the stock of anthropogenic heat in the air shed above our urban area. Once the steady-state stock of anthropogenic heat and the volume of the air shed are determined, the anthropogenic temperature can be calculated, because heat increase and temperature increase are related by the specific heat of the air. The stock of anthropogenic heat is equal to the flow of such heat into the air shed times the residence time of that heat. The flow is given in Eq. 4, above, and so only the residence time is needed to solve the problem.

The residence time can be estimated by calculating how long a warmed air parcel stays in the air shed. This period will depend on the wind speed and the size of the urban area. Our reasoning seems to be on the right track now, for the calculation explicitly involves the wind speed, the very parameter that determines urban heat dissipation.

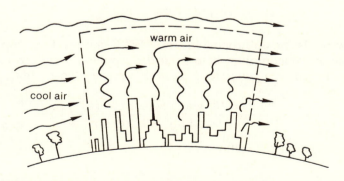

Figure III-7 The heat flux above a city. Cool air from the surrounding countryside replaces the rising hot air resulting from fuel combustion in the city. The urban air shed is shown as a box above the city.

Because convection carries the heat away from the urban area, certain details of this process must be specified. Assume that steady horizontal winds blowing at 10 km/hr sweep the air shed above the urban area. In reality, the speed and possibly direction of the wind depend on altitude and time, but ignore that for now in order to obtain an approximate answer.

Hot air generated at the surface will rise and be swept away by the horizontal wind. Since in order to leave the air space above the urban area an average parcel of air must be blown 10 km (one half the distance across the city[39]), the residence time of the anthropogenic heat is approximately 10 km/(10 km/hr) = 1 hr. Note that this estimate ignores the time required for the rising hot air to move vertically into the horizontal winds.

The stock of anthropogenic heat, H_a, within this air shed will thus equal the product of the residence time and the total inflow rate (power density × total area), or

$$H_a = (1 \text{ hr}) \times (280 \text{ W/m}^2) \times (20{,}000 \text{ m})^2 \qquad (5)$$

$$= 4.0 \times 10^{14} \text{ J.}$$

How great a temperature rise will be generated by this heat? To determine this we must estimate the mass of air that is to be warmed by the stored heat. (The more air there is to be heated, the smaller the effect of a given quantity of heat.) This mass equals the volume of the air shed above the urban area times the density of air. The volume of the air shed, in turn, depends on the average height to which vertical mixing of the anthropogenically heated air takes place. This altitude varies in time; during periods of atmospheric inversion it is low (less than 100 m), while on a clear day it can be 1,000 m or higher. Let us assume that, on the average, rising warm-air parcels occupy a box-shaped volume over the city with dimensions of 20 km × 20 km × 0.3 km.

The mass of air, M, in the air shed is therefore

$$M = (20 \text{ km})^2 \times (0.3 \text{ km}) \times (1{,}290 \text{ g/m}^3) \qquad (6)$$

$$= 1.5 \times 10^{14} \text{ g.}$$

---

39. This assumes the wind is blowing at right angles to one of the square edges of the urban area. Ambitious readers can show, using integral calculus, that the average distance the air parcel has to be blown is nearly independent of the orientation of the wind direction relative to the square-shaped urban area.

With these results in hand, the temperature increase, ΔT, can be estimated. It is given by the stock of heat, H_a, divided by the product of the mass of air, M, and the specific heat of air, C_p, or

$$\Delta T = \frac{H_a}{C_p M} = \frac{4.0 \times 10^{14} \text{ J}}{(1 \text{ J/g °C}) (1.5 \times 10^{14} \text{ g})}.$$

$$= 3 \text{ °C}.$$

(7)

This is a plausible result. Measurements in London and other metropolitan areas indicate values for T of approximately this magnitude.

How does the answer depend on the size of the urban area? If the area is a square of side b km, and within the area heat is generated at a fixed rate per unit area regardless of the value of b, then the residence time is proportional to b and the total rate of heat injection is proportional to b^2. Therefore, the stock of heat, H_a, is proportional to b^3. The mass of air in the air shed is proportional to b^2, however, and so the temperature will rise in proportion to $b^3/b^2 = b$. Of course, this is only true if the distance b is small enough that the mechanism of heat dissipation is wind dispersal. As the city grows in size, there comes a point when radiation losses exceed convective losses for the reasons discussed following Eq. 4. To include this size factor a more complicated combination approach must be taken, in which both heat dissipation mechanisms were included. But when b is only 20 km, the radiation term need not be considered. You can convince yourself of this by noting that when we did our calculations using radiation losses alone, the surface temperature increased by 71K.

If the urban area is longer in one direction than another, the residence time is proportional to the dimension of the urban area parallel to the wind direction. In this sense, residents of Boswash (the Boston to Washington seaboard) are lucky, for the prevailing winds are roughly perpendicular to the direction of sprawl. The same reasoning applies to air pollution buildup in urban areas. Planners of urban growth take heed.

A number of approximations were made in this derivation of ΔT. Many of the parameters were assigned numerical values that were averages of quantities that vary both stochastically (randomly) and in a periodic, predictable fashion. These parameters include the wind speed and direction, the height of the mixing volume, and the rate of inflow of anthropogenic heat. How does variation in a parameter about its average value affect the result of a calculation like this one?

Suppose, first, that a parameter, a, enters into a calculation in a linear fashion, so that the quantity, X, which is being calculated, depends on a as:

$$X(a) = ca,$$

(8)

where c is a constant of proportionality. Then, denoting the average of X and a by $\overline{X}$ and $\bar{a}$, it follows that

$$\overline{X} = X(\overline{a}). \tag{9}$$

In words, the average value of the function is equal to the function evaluated at the average value of its independent variable. Suppose, next, that the relation between X and a is concave upward, as shown in the upper panel of Figure III-8. Then, as that figure shows graphically,

$$\overline{X} > X(\overline{a}), \tag{10}$$

and thus the use of an average value of a parameter gives a result lower than the actual averaged value of the quantity of interest. By a similar argument, it can be shown that if the dependence of X on a is as shown in the lower panel of Figure III-8, then

$$\overline{X} < X(\overline{a}). \tag{11}$$

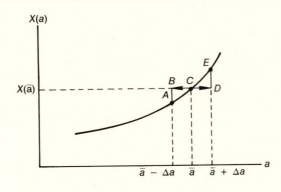

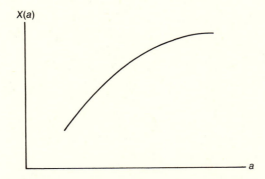

Figure III-8 The effect of fluctuations in an independent parameter, a, on the dependent variable, X. If X is a concave upward function of a, as shown in the top graph, then $X(a + \Delta a) - X(a) = ED$ is greater than $X(a) - X(a - \Delta a) = BA$. Hence, if a varies randomly between $a - \Delta a$ and $a + \Delta a$, then the average of X will exceed $X(\overline{a})$. The opposite is true for the bottom graph. (After Richard Levins, personal communication)

These simple and general results tell us something about the sign of the error made in using averaged values of the input parameters in any calculation. Consider, for example, the wind speed in this problem. Because the residence time is inversely proportional to the wind speed, v, and T is proportional to the residence time, it follows that the dependence of T on wind speed is concave and $\overline{\Delta T} > \Delta T(\bar{v})$. Thus the wind speed's variation about its average will cause the true averaged value of ΔT to be greater than the value we calculated. We have underestimated the value of ΔT by replacing the wind speed by its average value. It is the slower-than-average speeds that cause an increase in ΔT. The extent of that underestimation depends on the detailed probability distribution of wind speeds. You should now be able to figure out the sign of the effect of averaging over other parameters in this problem (such as the atmospheric mixing height or the rate of heat production) and in other problems throughout this book as well.

Another set of approximations, very different, was made in deriving Eq. 7. Only one mechanism of climate change, generation of waste heat, was considered. But cities can influence their climate in many ways, including (*a*) production of particles from fuel combustion, which can serve as cloud condensation nuclei and alter precipitation rates; (*b*) imposition of physical structures, such as skyscrapers, that block the wind and funnel it in sharp gusts at street corners; (*c*) the removal of trees and with them evapotranspiration; (*d*) the distribution of massive amounts of concrete, which alters the heat capacity of the surface; and (*e*) the alteration of surface albedo.[40]

EXERCISE 1: Derive Eq. 3 from Eqs. 12–14 in Problem III.6.

EXERCISE 2: How big does a square urban area have to be so that the increase in upward radiation from the surface equals the rate of convective loss of anthropogenic heat? Assume its inhabitants produce waste heat with a flux given by Eq. 4.

** *EXERCISE 3:* What qualitative effect on T_s will each of the factors listed in the last paragraph of the problem likely have? Construct a mathematical model and use it to estimate the numerical effect of one or more of these climate-changing alterations.

---

40. A thorough discussion of these and other effects, as well as a good list of references, can be found in Changnon (1976).

C.
Survival of Populations

In terms of conventional physics, the grouse represents
only a millionth of either the mass or the energy of an acre.
Yet subtract the grouse and the whole thing is dead.
—Aldo Leopold

The choices that human societies make influence whether populations, and even species, survive or die. Such choices include how
crop pests are dealt with, how toxic wastes are disposed of, whether
and how family planning is implemented, and how food is harvested.
The following problems explore some of the linkages between human
action and population survival. The populations discussed in some of
these problems are not human, but *Homo sapiens* is at risk in all cases.
Indeed, our fate is dependent on the fate of natural ecosystems, for
we depend upon such systems to provide a host of services that sustain life. Such services include maintaining the quality of air, water,
and soil; moderating the climate; and providing a genetic "library"
that we draw upon to develop crops and medicines.

The first two problems in this Section (III.10 and 11) concern populations that are deliberately managed for our potential direct benefit.
The first looks at the consequences of a chemical strategy for controlling unwanted species and the second at harvesting strategies for en-
hancement of wanted species. Problem III.12 discusses how populations can be poisoned inadvertently by a process called
biomagnification. This process led to the near eradication of the per-
egrine falcon in the United States in the days when DDT was widely
used there. Biomagnification of poisons in fish we eat can lead to our
own poisoning. Problem III.13 takes a look at some unfortunate rabbits. In Problem III.14 the changes now being set into motion in the

Chinese population are studied; China's recent commitment to controlling the growth of nearly one fourth of the population of *Homo sapiens* is a step of momentous global importance. To the final question (Problem III.15) I have no answer, but you will probably enjoy thinking about it.

10. Pesticides That Backfire

When a pesticide is applied to cropland, the target pest population sometimes increases while the predator population that had been controlling the pest declines. Explore this phenomenon with a model to determine whether it can be understood without very complicated equations.

· · · · · · ·

First let's see if we can understand in a qualitative way why the unintended effect might occur. If the pesticide attacks the pest population and reduces it, the predator population might be reduced as a result. That, in turn, would relieve grazing pressure on the pest population, allowing it to grow back. But then wouldn't the predator population recover as well? Certainly not if the pesticide attacks the predator as well as the prey. If it only attacks the prey, then recovery of the predator population might depend on whether prey killed by the pesticide become available as food for the predators. What do models tell us?

Probably the simplest model proposed to describe temporal changes of interacting predator and prey populations is the Lotka-Volterra model:

$$\frac{dX}{dt} = \beta XY - \alpha X \tag{1}$$

and

$$\frac{dY}{dt} = -\beta XY + \alpha'Y. \tag{2}$$

In these equations, X is the predator-population biomass and Y is the prey-population biomass. A constant, β, describes the strength of the predatory interaction leading to a biomass inflow (ingestion of prey) to the predator population ($+\beta XY$) and a corresponding outflow (loss of members) from the prey population ($-\beta XY$), while α is the per capita rate of outflow (for example, by excretion and death) from the predator population. Finally, α' is the per capita net input to the prey population (the difference between inputs from feeding and outflows other than by predation). Figure III-9 illustrates the stocks and flows in this model. The food supply of the prey population is not included in the model as an explicit variable, nor are other populations that are eaten by, or that eat, the predator. In addition, only the one population, X, is assumed to prey on Y. This simple model assumes, in effect, that other populations interacting with our predator and prey are constant, their effects included in the αX and $\alpha'Y$ terms.

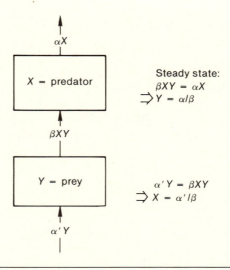

Figure III-9 The inflows and outflows in a simple two-level predator-prey model.

The steady-state solutions to these equations can be determined by setting $dX/dt = dY/dt = 0$. Our strategy will be to determine the steady-state solutions before and after a pesticide is applied. Denoting the steady-state solutions by $\overline{X}$ and $\overline{Y}$, Eqs. 1 and 2 yield:

$$\overline{X} = \frac{\alpha'}{\beta} \text{ and } \overline{Y} = \frac{\alpha}{\beta}.$$

If the direct effect of the pesticide is only to increase the death rate of the prey population, then the value of α' will decrease. This, in turn, will reduce the steady-state predator population, but it will do nothing, ultimately, to the steady-state prey population. If the pesticide increases the death rate of both predator and prey populations, then it will decrease the value of α' but increase the value of α. The effect of this is to decrease the steady-state predator population and increase the steady-state prey population. Note that in the first case, in which only the death rate of the prey is affected directly by the pesticide, the condition identified in the qualitative discussion at the beginning of the problem solution was satisfied. Namely, the effect of the pesticide as described by the model is to reduce the birth-minus-death rate of the pest without rendering the lost prey individuals available as food for the predator population.

EXERCISE 1: A number of ways to enhance the realism of the simple Lotka-Volterra model have been devised.[41]

41. May (1973) has written a delightful monograph describing a variety of approaches to modeling predator-prey relations in ecosystems. With great clarity he explains how general principles can be deduced from simple models.

One approach is to add to the equations a death-rate term that increases faster-than-linearly with the population size, thus simulating the effects of crowding on death rate. For example, the terms $-\gamma X^2$ and $-\gamma' Y^2$ might be added to the expressions for dX/dt and dY/dt, respectively.

Another approach to making the model more realistic adds a complication to the predator-prey interaction term that simulates an intuitively plausible effect: When the prey species gets very abundant, the predators are not limited by food supply and so the feeding-rate term is no longer proportional to the prey population. Mathematically, this is often represented by replacing the simpler βXY term with the Michaelis-Menten term $\beta XY/(1 + Y/K)$. As described by this more complicated expression, if the prey population is low, then the predation rate is proportional to the prey-species population, whereas if the prey population is high, then the predation rate approaches a value independent of the prey population.

Does the general conclusion observed in the simple model, namely the unintended consequence of increasing the death rate of the prey species, persist in models with either or both of the above-mentioned complications included?

11. **Optimal Harvesting**

You are harvesting from a wild or domestic population. What population parameters determine the maximum sustainable harvesting rate, and how does that rate depend on these parameters?

· · · · · · ·

The biomass of a population can be characterized by two rate constants—a gross loss or outflow rate, D, and a gross inflow rate, F. D is the sum of the death rate, the excretory and metabolic rates, and the emigration rate. F is the sum of all the rates at which biomass is added to the population by ingestion and immigration. The net growth rate, G, is given by

$$G = F - D. \tag{1}$$

The rate of change of the population's biomass, X, is

$$\frac{dX}{dt} = G. \tag{2}$$

Consider a preharvested population with the steady-state biomass X, characterized by a specified value of F and D, with $F = D$. An optimal, sustainable strategy would appear to be attained if the population could be harvested at a rate equal to the natural loss rate, D, so that the harvesting simply replaced the natural loss. Such a scheme appears sustainable because the harvested biomass is no greater than what would have exited the population anyway at the same rate had no harvesting occurred. It appears to be the maximum sustainable rate because if the harvest rate were to exceed D it would exceed F as well, rendering the net rate negative and sending the population into a decline.

However, there are several gaps in this reasoning. First, harvesting individuals just before they die naturally or emigrate and harvesting biomass equivalent to metabolic and excretory losses are highly impractical endeavors. If the population is harvested at the rate, D, some natural losses will occur anyway and the population will decline. Second, the gross rates, F and D, will generally depend on the population biomass, X. If X is harvested temporarily below its original steady-state value, then feedback effects will occur that will change the values of F and D. One such feedback effect might be the alleviation of overcrowding, leading to a decrease in the natural death rate and an increase in X. On the other hand, individuals in the less dense population might have trouble finding mates, and the species might drift toward extinction. Such effects could make our putative optimal rate less than or greater than optimal. With corrections leading in both directions, a model is needed to help sort out the situation.

Returning to the natural, unharvested population, let D be represented by two types of terms. One is a linear loss rate, αX, corresponding to a loss rate per unit of biomass, $\alpha X/X = \alpha$, independent of the size of the population. For some populations, such a population-independent per-unit-biomass rate is a very good approximation. For example, if the human population in your city or country is growing, the total death rate is also growing; but the population growth probably will not lead to an increase in the per capita death rate. This means that more people will die each year when the population is larger, but any individual's chances of dying in a given year remain unchanged. In a very crowded city, however, the per capita death rate might, in fact, increase as the city gets more crowded; diseases spread more effectively under crowded conditions, and murder and automobile accident rates might increase faster-than-linearly with population size.

In wild populations, the pressure of limited resources, including territory for breeding and food supply, can cause such a nonlinear effect, described by a second type of term in D. An individual's opportunity to obtain needed resources is diminished by large numbers of neighbors, and therefore the per capita death rate depends on population size.

This type of non-linear crowding effect is often represented mathematically by replacing $D = \alpha X$ with

$$D = \alpha X + \gamma X^2. \tag{3}$$

The αX term describes death and other losses by linear processes, and the γX^2 term corresponds to a loss rate per unit biomass of $\gamma X^2/X = \gamma X$. In the ecological literature the latter is called a density-dependent effect. The specific density-dependent behavior in Eq. 3 is called the Verhulst effect.

The gross growth rate, F, is most simply taken to be proportional to X_i, corresponding to a constant gross growth rate per unit of biomass:

$$F = \sigma X. \tag{4}$$

A modification that is sometimes made replaces this term with

$$F = \frac{\sigma X^2}{X_0 + X}, \tag{5}$$

where X_0 represents a critical level of biomass. When X is very large compared to X_0, this expression behaves the way Eq. 4 does, for in that case

$$F \approx \frac{\sigma X^2}{X} = \sigma X. \tag{6}$$

But when X is small compared to X_0, the F in Eq. 4 goes to zero faster than the F in Eq. 5. A type of density-dependent behavior called the Allee effect is described in Eq. 5, which represents mathematically the feedback mechanism that can cause a depressed population to drift toward extinction because of reproductive difficulties. We will use the simpler expression in Eq. 4 here, but see Exercise 3.

A useful model of the unharvested population, from which we can then explore how harvesting strategies relate to population parameters, combines Eqs. 2, 3, and 4:

$$\frac{dX}{dt} = \sigma X - \alpha X - \gamma X^2$$
$$= rX - \gamma X^2, \tag{7}$$

where $r = \sigma - \alpha$. This is customarily written in the form

$$\frac{dX}{dt} = rX \left(1 - \frac{X}{K} \right), \tag{8}$$

where $K = r/\gamma$. Eq. 8 is sometimes referred to as the logistic equation.

K is called the carrying capacity; it is the maximum value of X that can occur without causing a population decline. It is a stable point in the sense that if X starts out below K, then dX/dt will be positive and the population will increase to K; if X exceeds K, then $dx/dt < 0$ and the population will decline to the value of K. Note, for later reference, that the growth rate, dX/dt, is maximum where $X = K/2$ (see Figure III-10).

Now suppose that harvesting begins. The harvesting process must fall into one of two broad categories. Either the harvest rate is fixed at a constant value independent of the population size, or it can vary. If it can vary, it is most likely set proportional to the population size. In the first category, a fixed yield is established. Where each of a fixed number of hunters is allowed a quota of kills each year, for example, the total harvest rate is independent of the population size. To achieve their quotas in years when the prey population is depressed, the hunters may have to hunt many extra hours.

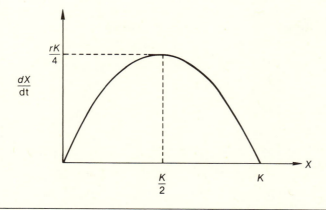

Figure III-10 The relation between the net growth rate of the population, dX/dt, and the size of the population, X, in a logistic-equation model of population growth. K is the carrying capacity and r is the linear growth rate constant.

The second category involves harvesting at a fixed effort. For example, if a fixed number of fishermen with specified equipment ply the ocean waters for a certain number of hours per year, then the catch is likely to depend on the population size. A reasonable first approximation assumes the harvest rate is linearly proportional to the population, with the proportionality constant equal to a measure of the harvesting effort. With these harvesting strategies in place, the population equation becomes

$$\frac{dX}{dt} = rX\left(1 - \frac{X}{K}\right) - h \tag{9}$$

in the first case, and

$$\frac{dX}{dt} = rX\left(1 - \frac{X}{K}\right) - EX \tag{10}$$

in the second. In Eq. 9, h is the fixed harvest rate, while in Eq. 10 E is the fixed effort expended in harvesting.

Consider Eq. 10. What level of effort, E, corresponds to the maximum sustainable yield? We will assume that the effort is constant rather than time-varying. In that case, the population eventually reaches a steady state, $X = \overline{X}$, and thus the harvest is indeed sustainable. The value of $\overline{X}$ is given by setting $dX/dt = 0$, or

$$E\overline{X} = r\overline{X}\left(1 - \frac{\overline{X}}{K}\right). \tag{11}$$

Solving for $\overline{X}$,

$$\overline{X} = \left(1 - \frac{E}{r}\right)K. \tag{12}$$

Hence the sustainable harvest rate is

$$E\overline{X} = E\left(1 - \frac{E}{r}\right)K. \tag{13}$$

The maximum sustainable yield will be obtained when $E\overline{X}$ is maximized. This occurs at a value $E = E_m$, given by

$$\frac{d}{dE}\left[E\left(1 - \frac{E}{r}\right)K\right] = 0 \tag{14}$$

or

$$E_m = \frac{r}{2}. \tag{15}$$

Substituting Eq. 15 into Eqs. 12 and 13, we find the population under optimum harvesting, $\overline{X}_m$, to be

$$\overline{X}_m = \frac{K}{2}, \tag{16}$$

and the optimal harvesting rate is

$$E_m\overline{X}_m = \frac{rK}{4}. \tag{17}$$

Eqs. 16 and 17 can be interpreted as follows. In the absence of harvesting ($E = 0$), the net growth rate, dX/dt, is maximum when $X = K/2$ (see Figure III-10). At this value of X, $dX/dt = rK/4$. Thus, the optimal harvesting rate equals the maximum net growth rate of the unharvested population, and the size of the sustainable population under optimal harvesting equals the size of the population that yields that maximum net growth rate in the unharvested population. This is all quite reasonable. Less obvious is the fact that the optimal level of effort, E_m, which is the quantity of concern to the harvester, is equal to $r/2$ or half the linear growth rate of the unharvested population.

Let's now shift our perspective and explore some economic considerations. Suppose the harvested population is viewed as a commodity that fetches a certain price, p, per unit of yield. The gross rate of income, I, is then,

$$I = pEX, \tag{18}$$

where EX is, again, the harvest rate. Suppose, also, that the rate of expense, C, to the harvester is proportional to the effort, E, expended in harvesting. Thus

$$C = cE. \tag{19}$$

Then the total rate of net income or profit, Y, is

$$Y = I - C \tag{20}$$
$$= pEX - cE.$$

Let us maximize the rate of net income, keeping the sustainability criterion that $dX/dt = 0$.

Since $dX/dt = 0$, we must still enforce Eq. 11. Therefore, $X = \overline{X} = (1 - E/r)K$, and

$$Y = E[p(1 - E/r)K - c]. \tag{21}$$

Maximizing Y with respect to E yields

$$\frac{dY}{dE} = pK - 2E\,\frac{pK}{r} - c = 0 \tag{22}$$

or

$$E_m = \frac{rpK - cr}{2pK} \tag{23}$$
$$= \frac{r}{2} - \frac{cr}{2pK}.$$

The value of X at the level of maximum sustainable net income is now obtained by substituting Eq. 23 into Eq. 12:

$$\overline{X} = K\left[1 - \frac{1}{r}\left(\frac{r}{2} - \frac{cr}{2pK}\right)\right] \tag{24}$$
$$= \frac{K}{2} + \frac{c}{2p}.$$

So we see that profit maximization results in a higher steady-state population (Eq. 24) than does yield maximization (Eq. 16). Comparing Eq. 23 with Eq. 15, we see that profit maximization results in a lower effort rate. How does the actual yield, $E_m\overline{X}_m$, compare in the two cases? Profit maximization gives

$$E_m\overline{X}_m = \left(\frac{r}{2} - \frac{cr}{2pK}\right)\left(\frac{K}{2} + \frac{c}{2p}\right) = \frac{rK}{4} - \frac{c^2r}{4p^2K}, \tag{25}$$

which is less than the value $rK/4$ obtained for the yield-maximization case.

The effect on the steady-state population of bringing in economic considerations in this manner is to optimize at a higher value of X. This is a good thing if one's interest is in species preservation. Larger wild populations can sustain higher genetic diversity than smaller ones. Because of this and because smaller populations are more vulnerable to inevitable fluctuations in their numbers, larger populations are usually less vulnerable to extinction if a sudden change in their habitat occurs, or if harvesting pressure increases rapidly.

The consequences for a population's survival of the complex interplay of genetic diversity, changes in predation pressure, habitat alteration, and fluctuations in numbers are difficult to incorporate realistically in models. When working with simple models, like our Eq. 10, we should not lose sight of these complex effects. The optimal-profit and optimal-yield strategies both involve steady-state populations that theoretically should persist forever, in the narrow context of the model. Nonetheless, qualitative considerations inform us that the population is safer under an optimal-profit strategy.

A more complete analysis of economic factors influencing harvesting rate and optimal population size leads to a bleaker picture, however. Our analysis did not include a discount rate, which is a measure of the rate at which money can grow in real value (after inflation) if invested. People would rather get a dollar this year than next, and therefore from this narrow perspective an individual harvested from a population is worth more this year than next. With the discount factor incorporated into the analysis, it can turn out, in some cases,[42] that $\overline{X}_m < K/2$. This can bode ill for the population being harvested.[43]

EXERCISE 1: Show that in the optimal-yield case for the model described by Eq. 10, the natural loss rate, $\alpha X + \gamma X^2$, exceeds the optimal harvest rate.

EXERCISE 2: Repeat the analysis presented here that was based on Eq. 10, but assume a constant harvest rate, h, as in Eq. 9.

* *EXERCISE 3:* Repeat the analysis using a harvest rate, EX, as in Eq. 10, but with the Allee effect included in the population model.

42. This is discussed in an article by Clark (1977). The concepts and notation used here borrow heavily from Clark's treatment. Our treatment, here, of optimal harvesting is limited to the harvesting of a single species. For a discussion of harvesting strategies appropriate to multispecies fisheries, see May et al. (1979).
43. A lucid, highly knowledgeable treatment of the causes and consequences of species extinction can be found in Ehrlich and Ehrlich (1981).

12. Biomagnification of Trace Substances

How does biomagnification of a trace substance occur? Specifically, identify the critical ecological and chemical parameters determining bioconcentrations in a food chain. In terms of these parameters, derive a formula for the concentration of a trace substance in each link of a food chain.

· · · · · · ·

Imagine a young fish growing up in a lake and eating nothing but plankton. As the fish grows from, say, 10 g at birth to its average weight at death of 10^3 g, all of its newly acquired flesh and bone will be derived from the plankton it eats. To add on 990 g of tissue, it will have to eat far more than 990 g of plankton, because of excretory and metabolic losses. Excretory losses, in the sense used here, include the sloughing off of old tissue as it is replaced with new. The typical growing fish eats 10 g of plankton to grow by about 1 g of body weight. In the ecological literature this factor of 10% (the ratio of the weight it gains to the weight of the food it eats) is called an *incorporation efficiency*. The remaining 90% of the food the fish eats is excreted or metabolized.

Suppose, now, that the plankton contains a trace substance, such as DDT or mercury. If a greater proportion of the trace substance than of the plankton is retained in the fish rather than excreted and metabolized, then the concentration of the trace substance in the animal will build up to a level greater than that in the plankton. Similarly, if a population of osprey prey upon the fish, they too may preferentially retain the trace substance in their food; over the lifetime of the osprey, the concentration of the substance can build up to an even greater level than that in the fish. The fraction of ingested trace substance retained by an organism is called the retention factor.

From this qualitative discussion, the factors that determine the degree of concentration of a trace substance in an organism can be deduced:

(1) Preferential Retention of the Substance in the Body In the extreme case, the rate of bioconcentration is greatest when all the trace substance an animal ingests is stored in body tissue and none is metabolized[44] or excreted.

(2) Fraction of Ingested Food Incorporated into New Tissue If an animal is very inefficient in building new tissue out of its food source, it must consume a lot of food to grow by any specified amount. That

---

44. Some substances, such as alcohol, are metabolized so rapidly by many organisms that bioconcentration is not a problem.

larger amount of ingested food is accompanied by a larger amount of trace substance. Thus, for a given retention factor, the lower the incorporation efficiency, the greater the rate of bioconcentration.[45]

(3) Ratio of Weight at Death to Weight at Birth   For a given retention factor and incorporation efficiency, the more weight an animal puts on during its lifetime, compared to its weight at birth, the greater is the percentage increase in the concentration of a trace substance over the lifetime of the organism. Note that while this ratio, the "relative growth factor," is sometimes correlated with the lifetime of the organism, it need not be. Fish generally don't live as long as humans, but their proportional weight gain is much larger. If an organism is fully grown well before death (like humans or birds, but not fish or trees), and if that organism lives many years in its fully grown state, then during these mature years its intake of the trace substance continues but its body weight does not change. Therefore, the longer the period in which the organism is fully grown, the greater the concentration of the trace substance in the organism at death. However, that effect is already described by the incorporation efficiency, which, if given as an average over the lifetime of the organism, will reflect a possible lack of growth during later years.

(4) Location in the Food Chain   The osprey bioconcentrates the trace substance from the fish it eats and, if other factors are equal, accumulates a higher concentration than that in the fish. The higher in a food chain an organism feeds, the greater is the concentration effect for that organism.

(5) Environmental Contamination   The concentration of the trace substance in the plankton reflects that in the water. The contamination of soil and water initiates the food-chain effect and its amount is therefore an important determinant of the ultimate concentration in all organisms.

Now that the qualitative picture is established, let's build a plausible mathematical model to determine the quantitative importance of the various factors described above. Consider N populations, each in a steady state, whose feeding pattern is described by a linear food chain.[46] Ignore immigration and emigration. Let $X_1, \ldots, X_N$, be the biomasses of the populations, where X_j eats X_i if $j = i + 1$. None of the N populations feed on X_N; dead individuals from that population are decomposed by organisms not included among the N populations here.

---

45.   This discussion assumes that the trace substance is ingested along with the food used by the organism. Some trace substances, however, can be incorporated into body tissue even if the substances they are found in have no food value and the rate of intake of trace substance is independent of the rate of food ingestion. For such substances, the concept of incorporation efficiency is not relevant to the problem at hand.
46.   This is, of course, an abstraction. Real populations rarely line up neatly in a simple feeding *hierarchy*. Rather, a food *web* is more typical.

Assume that D_i is the total rate at which biomass exits from the ith population. This total rate is the sum of metabolism plus excretion, e_i, predation, p_i, and other causes of death, d_i:

$$D_i = e_i + p_i + d_i. \tag{1}$$

The total inflow rate is F_i, consisting entirely of ingested food.[47] Note that birth is not an inflow; the process of birth adds no biomass to the population, but only divides the existing biomass into smaller pieces. The steady-state assumption is equivalent to

$$F_i = D_i \tag{2}$$

for $i = 1, \ldots, N$. All rate constants, D_i, e_i, p_i, d_i, and F_i are in units of biomass per unit time. The predation from level i is the input to level $i + 1$, and so

$$p_i = F_{i+1}. \tag{3}$$

The rate that ingested food is not metabolized or excreted is $p_i + d_i$, and therefore the incorporation efficiency, E_i, for each population is

$$E_i = \frac{p_i + d_i}{F_i}. \tag{4}$$

Figure III-11 illustrates the stocks and flows in our model.

At time $t = 0$, assume a trace substance is added to the environment (e.g., the water or soil in which X_1 grows) so that a constant environmental contamination, characterized by a concentration of C_0, is maintained for all time thereafter. Denote the *concentration* of pollutant at time t in each population by $C_i(t)$. C_i has units of mass of trace substance per unit biomass. The *amount* of trace substance in the ith population at time t is $M_i(t) = X_i C_i(t)$.

Define A_i and B_i to be the inflow and outflow of the trace substance (in units of mass per unit time) to each population. The inflow rate, A_i, to the ith population is determined by the predation rate of i on $i - 1$ and by the concentration of trace substance in the $(i - 1)^{\text{th}}$ population, C_{i-1}. In particular,

$$A_i = p_{i-1}C_{i-1}. \tag{5}$$

The rate of loss of the trace substance from the ith population can be expressed as a sum of two terms. The first equals the concentration, C_i, times the rate of outflow (by predation or other forms of death)

---

47. If the $i = 1$ population is a photosynthesizer, then F_1 should be thought of as "ingested" CO_2, H_2O, inorganic nitrogen, and other nutrients.

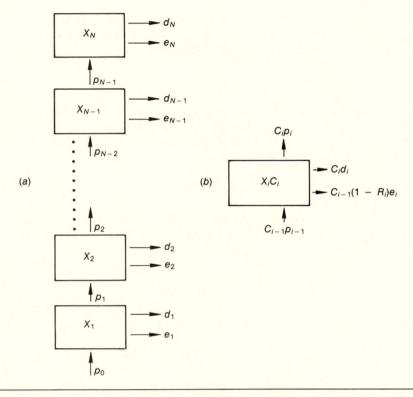

Figure III-11 (a) The inflows and outflows of biomass through the N levels of a food chain. X_i is the biomass of each level, p_{i-1} is the inflow to the ith level by predation, while e_i and d_i are outflows resulting from metabolism plus excretion (e_i) and death (d_i). (b) The inflows and outflows of a trace substance for the ith level. The concentration of the trace substance is C_i. R is a retention factor (see text).

of trace-substance-containing biomass from that population. The second is proportional to C_{i-1} times the metabolism-plus-excretion rate,[48] with the proportionality constant equal to, or a fraction less than, one. That fraction is one minus a retention factor, R_i, where R_i is the fraction of ingested trace substance retained in the body. In equation form, the outflow rate is

$$B_i = C_i(p_i + d_i) + C_{i-1}(1 - R_i)e_i. \tag{6}$$

Eq. 6 states that predation and other types of death ($p_i + d_i$) lead to a loss of trace substance from the population equal to 100% of whatever is in the bodies of dying organisms; it also states that metabolism and excretion lead to the loss of only a fraction, $(1 - R_i)$, where R_i is the retention factor for the ith population (see Figure III-11).

---

48. C_{i-1}, not C_i, enters into the loss rate here because excrement has a concentration of trace substance governed by that of the food rather than that of the body as a whole.

The steady-state conditions on the M_i (or[49] C_i), determined by setting $A_i = B_i$, are:

$$p_{i-1}C_{i-1} = C_i(p_i + d_i) + C_{i-1}(1 - R_i)(e_i). \tag{7}$$

Thus the ratio, C_i/C_{i-1}, in the steady state, is

$$\frac{C_i}{C_{i-1}} = \frac{p_{i-1} - (1 - R_i)e_i}{p_i + d_i}. \tag{8}$$

Using Eqs. 1–4, this can be rewritten

$$\frac{C_i}{C_{i-1}} = 1 + R_i\frac{(1 - E_i)}{E_i}, \tag{9}$$

where E_i is the incorporation efficiency of the ith population.

Eq. 8 can be applied even to the case $i = 1$ if p_0 refers to the rate of uptake of the contaminated environmental medium by the population, X_1, at the bottom end of the food chain. By repeated iteration of Eq. 9, the steady-state concentration of trace substance in each population turns out to be:

$$C_1 = C_0\left[1 + R_1\frac{(1 - E_1)}{E_1}\right]$$

$$C_2 = C_0\left[1 + R_1\frac{(1 - E_1)}{E_1}\right]\left[1 + R_2\frac{(1 - E_2)}{E_2}\right]$$

$$\begin{matrix} \cdot \\ \cdot \\ \cdot \end{matrix} \tag{10}$$

Each factor $[1 + R_i(1 - E_i)/E_i]$ is greater than or equal to 1. This follows from the fact that R_i and E_i are both non-negative and $E_i < 1$. Hence $C_i > C_j$ if $i > j$. For fixed E_i and C_{i-1}, the larger the retention factor, R_i, the greater is C_i. For fixed R_i and C_{i-1}, the smaller the incorporation efficiency, E_i, the greater is C_i. And the larger the environmental concentration, C_0, the larger are all the C_i. Thus, four of the five qualitative conditions we deduced earlier are seen to fit simply into place from the model. However, the third condition (ratio of weight at death to weight at birth) is not reflected in any way in Eq. 10. (See Exercise 3).

In deriving Eqs. 10, we assumed that the environmental concentration, C_0, was maintained at a constant value. We then solved for the steady-state values of the C_i. Moreover, the quantities of toxic substance lost from the organisms by excretion and death (e_i and d_i) did

---

49. Note that if the M_i are in steady state, so are the C_i. This follows from the steady-state assumption for the X_i.

not return to the environment to be "recycled." Exercise 2 offers an opportunity to examine that complication, with the environment now an active rather than a passive link in the food chain.

EXERCISE 1:   Derive Eq. 9 from Eqs. 1–4 and Eq. 8.

* EXERCISE 2:   Suppose the trace substance is a conserved quantity, in the sense that $M_0 + M_1 + \ldots + M_N$ equals a constant. Now the losses of trace substance from the N populations—the flows $C_i d_i$, $C_{i-1}(1 - R_i)e_i$, and $C_N P_N$—are recycled back to the environmental compartment, adding to the value of C_0. Write a plausible set of equations analogous to Eqs. 5 and 6 to describe this situation, and solve for the steady-state value of the C_i.

* EXERCISE 3:   A reasonable guess is that the third qualitative condition, the ratio of weight at death to weight at birth, is a determining factor in the magnitude of the increase of trace substance in an *individual* during its lifetime but determines neither the actual steady-state values of C_i for the populations nor the rate at which these build up to their steady-state values in the population over time. Construct a simple model for the change in concentration of a trace substance in an individual whose food has trace substance in it, and determine how the weight-ratio effect enters into the estimate of the increase in concentration from birth ($t = 0$) to death ($t = T$). To tackle this problem, assume the individual's growth rate is constant in time from birth to death and write a pair of equations for dw/dt and $d(wb)/dt$, where w is the biomass of the individual and b is the concentration of trace substance in the individual. Assume $b(0) = 0$, and show by integration of these equations that the concentration of trace substance in the individual at death, $b(T)$, is given by

$$b(T) = a\left[1 - \frac{w(0)}{w(T)}\right].$$

Here, a is a constant given by

$$a = \frac{c(p_0 - e_0 + e_0 R_0)}{p_0 - e_0},$$

where p_0 is the ingestion rate, e_0 is the metabolism-plus-excretion rate, R_0 is the retention coefficient for the individual, and c is the concentration of trace substance in the individual's food.

** EXERCISE 4:   Consider a food chain in which an environmental contaminant is in steady state as described by Eqs. 10. Suppose that at $t = 0$ the environment is "cleaned" so that $C_0 = 0$ thereafter. How will the C_i subsequently depend on time, and how long will it take for each population to lose 50% of its steady-state concentration of the contaminant? Qualitatively, what properties of the population govern this time constant?

13. Rabbits on the Road

Recently driving across Nevada I counted 74 dead but still easily recognizable jackrabbits on a 200-km stretch of Highway 50. Along the same stretch of highway, 28 vehicles passed me going the opposite way. What is the approximate density of the rabbit population to which the killed ones had belonged?

· · · · · · ·

This is not a task for those of you who are squeamish about making assumptions. It's a problem best approached in the spirit of the cobbler problem (I.1).

Let the highway kill rate for rabbits be K, in units of number of rabbits per day. Let R be the size of the population of rabbits to which the killed ones belonged. Thus R is the population at risk—a somewhat fuzzy notion that we'll make more precise later. We now have a flow, K, and a stock, R. The ratio of R to K is the residence time associated with the particular exit pathway of being run over. The inverse of this time constant is a probability per unit time, p, that a rabbit will get run over, or

$$p = \frac{K}{R}.\tag{1}$$

It is easy to see why p, the inverse of the residence time, is a probability per unit time: K is the number of rabbits per unit time that get run over and R is the total population; therefore, K/R is the fraction of the population per unit time that gets run over. Such a fraction is just what one means by a probability per unit time.

That same probability can also be calculated from two other pieces of information we may be able to estimate: the average number, n, of road-crossings per unit time attempted by each rabbit; and the likelihood, r, that if a rabbit attempts to cross the road it will be run over. In particular,

$$p = nr.\tag{2}$$

Combining Eqs. 1 and 2,

$$R = \frac{K}{nr}.\tag{3}$$

R is the population at risk. We can assume that this population inhabits the land on both sides of the highway along the entire

200 km stretch. The population density, d, is then the population along that stretch divided by the area, A, that the vulnerable rabbits might reasonably be expected to inhabit:

$$d = \frac{R}{A}. \tag{4}$$

The land area, A, inhabited by the population whose unlucky members were censused, is 200 km long and of a width we will call w in units of km. Thus,

$$A = 200w \ (km)^2 \tag{5}$$

and

$$d = \frac{K}{200wnr} \frac{rabbits}{(km)^2}. \tag{6}$$

To determine K, we must estimate the period during which 74 rabbits were run over. If dead rabbits remain on the road in recognizable condition for many days, the observed corpses correspond to a lower death rate than if they were all killed within the previous day. Since large numbers of ravens and vultures enjoying "lunch on the road" could be observed doing their efficient job of removing corpses, let's assume that the 74 rabbits were killed in one day. Using round numbers,

$$K = 10^2 \ rabbits/day. \tag{7}$$

Estimating the number, n, of crossings per unit time attempted by each rabbit is our most difficult task. Suppose the rabbits hop randomly about their habitat for 1 hr/day. If rabbits hop at a speed of s km/day, then they will cover about $0.04s$ km per day and, very roughly, will cross the road $0.04 \ s/w$ times per day. Therefore,

$$n = \frac{0.04s}{w} \ (days)^{-1}, \tag{8}$$

with s in units of km/day and w in units of km. Therefore, substituting Eqs. 7 and 8 into Eq. 6,

$$d = \frac{10^2}{200w\left(\dfrac{0.04s}{w}\right)r} \frac{rabbits}{(km)^2}. \tag{9}$$

The likelihood that a rabbit attempting to cross the road will be killed, denoted by r, is estimated from knowledge of the time it takes to cross the road and the time between vehicles. If the "lethal" part of the road (one lane in each direction) is assumed to be 5 m (or 0.005 km) wide, then crossing takes a time period of $0.005/s$ in units of days. During the time a rabbit takes to cross the road a car or truck will travel a distance equal to the speed of the vehicle times the time interval, $0.005/s$. Assuming an average automobile speed of 100 km/hr (here I am advocating the use of round numbers, not the breaking of the speed limit), then the vehicle will travel $12/s$ km while the rabbit is crossing. If the vehicle is within that $12/s$ km stretch of road at the time a rabbit begins its crossing, the rabbit will be killed.

At any given location along the highway, the odds that a vehicle will be within this critical distance of $12/s$ km depend on the density of automobiles. Knowing that 28 vehicles passed me in 2 hr from the opposite direction, I can deduce that if I had been standing still at the roadside for those two hours, only half that many, or 14, would have passed me from that same direction. This is equivalent to 7/hr, and so the vehicles in one direction are about 0.14 hr or 14 km apart.

The value of r is, very roughly, the ratio of the critical distance $12/s$, to the actual average interval, 14 km, or[50]

$$r = \frac{12}{14s} \approx \frac{1}{s}. \tag{10}$$

Despite appearances, r is dimensionless, being a probability equal to the ratio of two distances. Substituting into Eq. 9, the population density is given by

$$d \approx 10 \text{ rabbits/km}^2. \tag{11}$$

Note that our answer is independent of s and w, both of which canceled out along the way.

EXERCISE 1: What are the unstated assumptions made in deriving Eq. 1?

EXERCISE 2: The net primary productivity of desert habitat like that in Nevada is (see Appendix XII.2) about 30 grams of carbon per square meter per year. Is this adequate to support the rabbit density given by Eq. 11?

---

50. The probability of getting killed is one minus the probability of crossing safely, and the probability of crossing each lane safely is $1 - (6/14s)$. Therefore, the probability of crossing both lanes safely is $[1 - (6/14s)]^2$ and $r = 1 - [1 - (6/14s)]^2$. This is approximately $12/14s$ if $6/14s$ is small compared to unity.

EXERCISE 3:   The probability that a rabbit crosses the road n times safely and then gets run over on crossing $n + 1$ is $r(1 - r)^n$. The mean number of crossings, $\bar{n}$, before a rabbit gets run over is then

$$\bar{n} = \sum_{n=0}^{\infty} nr(1 - r)^{n-1}.$$

Evaluate this expression and then show that if $r << 1$, $(1 - r)^{\bar{n}} = e^{-1}$. [Hints:

$$\sum_{n=0}^{\infty} a^n = \frac{1}{1 - a}$$

and

$$\sum_{n=0}^{\infty} na^n = a\frac{d}{da}\sum_{n=0}^{\infty} a^n = \frac{a}{(1 - a)^2}$$

if the absolute value of a is less than 1. In the limit as r approaches zero, $(1 - r)^{1/r}$ approaches e^{-1}.]

* *EXERCISE 4:*   Estimate in two ways the average lifetime of a rabbit that gets run over. First use the residence-time approach. Here you will have to estimate the width, w, of the 200 km long habitat.

A second approach begins with the following argument. The probability of a rabbit's not getting run over when it crosses the road is $1 - r$. The probability that it will cross twice without getting run over is $(1 - r)^2$. When a rabbit has made N crossings, where N is such that $(1 - r)^N = e^{-1}$, a rabbit can be said to have "used up its luck" (see Exercise 3). In other words, its average lifetime will be roughly equal to N times the time interval between crossings. That time interval can be estimated from Eq. 8 if values for s and w are first guessed. Using $s = 50$ km/day and $w = 10$ km, calculate the lifetime both ways.

EXERCISE 5:   Next time you take a long automobile trip, try to estimate how many rabbits, turtles, or birds (or whatever other animals you see slaughtered on the highway) are run over every year in the part of the world you are travelling in.

14. Approaching a Steady-State Population in China

The Chinese population in 1980 was about 10^9. Shortly thereafter, China set forth the goal of reducing its population to 7×10^8. To accomplish this, China is strongly encouraging one-child families. How long will it take China to achieve this goal under the following assumptions:

1. Sex-specific differences in the death rates and age distribution of the population can be ignored.

2. The age-specific death rates do not change.

3. A one-child-per-family policy begins at the start of 1980. In each decade from 1980 onward, until the population reaches 7×10^8, half the women who were in the age group 10–19 at the start of the decade, and half who were in the age group 20–29, have a child.

4. At the start of 1980, the age distribution and age-specific death rates (deaths per year per thousand of the population of specified age) are:[51]

Age Group	Number ($\times 10^6$)	Age-Specific Death Rate
0–9	235	7.90
10–19	224	0.98
20–29	182	1.60
30–39	124	2.50
40–49	95	4.80
50–59	69	11.00
60–69	42	27.00
70–79	24	63.00
> 80	6	145.00

· · · · · · ·

To make a rough guess, note first that the present (1980) average death rate, weighted by the populations in each ten-year age category, is

$$D = \frac{7.9\,(235) + 0.98(224) + \ldots + 145(6)}{235 + 224 + \ldots + 6}$$

$$= 7.4 \text{ deaths per year per } 10^3 \text{ population.}$$

(1)

This is a 0.74% per year loss rate for the population. Next, we have to estimate the birth rate under a one-child-per-family policy. If every

---

51. The age-distribution data were calculated from birth rates, death rates, and age distribution data for 1975 given in Bannister and Preston (1981). An apparent decline in the age-specific death rates between 1975 and 1980 is ignored here.

pair of people had two children, not one, in their lifetime (which we take to be about 71 years—see Exercise 5), then they would just replace themselves. Very roughly, the birth rate would be about 1,000 per 71 years per 1,000 people, or a gross addition of about 14 per year per 10^3 population. With a one-child-per-family policy, the birth rate would be half of this, or 7 per year per 10^3. This is a 0.7% per year birth rate. Combining this with the death rate gives a net birth-minus-death rate of -0.04% per year. (The relative uncertainty in this estimate is huge because 0.04% is the small difference between two larger numbers that are both uncertain. But as the argument below shows, there is something even more fundamentally wrong with tackling the problem this way.) To proceed, $N(t) = N(0)\,e^{-0.0004t}$, with t evaluated in years.[52] If $N(0) = 10^9$, then $N(t) = 7 \times 10^8$ at a time $t = T$, where

$$-0.0004T = \log_e (0.7) = -0.36 \qquad (2)$$

or

$$T = 900 \text{ yr.} \qquad (3)$$

Fortunately, this Methuselean estimate is way off. In fact, given our assumptions, the goal would be reached within the lifetime of a young Chinese today.

You can tell this first approach is way off by taking the limiting case of a no-children-per-family policy. Then the birth rate would be zero and the death rate 0.74% per year. This would imply $N(t) = N(0)\,e^{-0.0074t}$. At the end of, say, 200 years, N would equal 2.3×10^8. However, the population couldn't possibly equal 2.3×10^7 after 200 years of a no-birth policy, unless many individuals lived 200 years.

The problem with the guesswork above is that it ignores the changing age distribution of the population subsequent to the implementation of a new demographic policy. To include this complexity, we have to set up a proper box model.

Each 10-year age group of the Chinese population can be considered a box. For the youngest age group (0–9), the inflow to the box is by birth and the outflow is either by death or by aging to the next oldest box (10–19). For the 10–19 through 70–79 boxes, inflow is by aging from the next youngest box and outflow is either by death or by aging to the next oldest box. For the oldest age-group box (> 80), inflow is by aging and outflow is by death.

For convenience, we will represent the age distribution of China's population (in units of 10^6 people) with a column of numbers (called

---

52. We can use the annual net growth rate as an instantaneous rate because it is very small and thus $\exp[-0.0004]$ is nearly identical to $1 - 0.0004$ (see Problem I.5).

a vector). We denote the vector by $\mathbf{P}(t)$. The 1980 value of $\mathbf{P}$, in units of 10^6 people, is given in the problem statement.

$$\mathbf{P}(1980) = \begin{vmatrix} 235 \\ 224 \\ 182 \\ 124 \\ 95 \\ 69 \\ 42 \\ 24 \\ 6 \end{vmatrix}. \tag{4}$$

The sum of the numbers is 1,001, and so the total population is 1.001×10^9.

If there were no births or deaths, $\mathbf{P}(1990)$ would be easily derivable from $\mathbf{P}(1980)$ alone. The first entry in $\mathbf{P}(1990)$ would be zero, the second would be 235 and each subsequent one would just be shifted down one ten-year interval from the 1980 value, except for the last entry, which would be $6 + 24 = 30$. However, births and deaths have to be taken into account. To calculate the number of births in the ten-year period from 1980 to the end of 1989, we have to use assumptions 1. and 2. Assumption 1. implies that half the people in any age group are female. At the start of the decade 1980–1989, half the number of females in the 10–19 age group was $224/4 = 56$ (in units of 10^6) and half the number of females in the 20–29 age group was $182/4 = 45.5$. Therefore, by assumption 3. the number of births in that decade was 101.5×10^6.

The astute reader will realize at this point that our assumption 3. is not exactly equivalent to a one-child-per-family policy. For one thing, we have ignored the possibility of twinning. In addition, we have not specified the age-dependent probability of whether females who die during the childbearing years (roughly 20–29, given our assumptions) first have a child; nor have we specified whether the parents of a child who dies soon after birth have a "replacement" child. Because death rates can vary greatly with age within a ten-year age bracket and age distributions can change over a ten-year period, it is not possible to guarantee that every female who lives to a specified year has a child surviving to a specified year. We would like to work with death rates and childbearing rates that were continuous functions of age; instead of leaping forward in ten-year jumps, we could then calculate the smooth time-evolution of the population. However, such data will never exist for real populations.

To proceed, we will ignore the possibility of twinning. We will also assume that at the end of a decade, the number of survivors in the

0–9 age group is equal to the number of births calculated by assumption 3. In effect, we thus assume that some replacement of dead children by second births occurs. The top entry in **P**(*1990*) is then 101.5.

The second entry will be 235 minus the number of deaths (in units of 10^6) that occur during the decade 1980–1989 among the collection of people (called a cohort) who were in the 0–9 age group in 1980. Members of this cohort will span the ages of 0–19 during the decade, and therefore it is a reasonable approximation[53] to take the effective death rate to be the average of that rate for ages 0–9 and 10–19. The number of deaths within the cohort will equal, approximately, the product of the average size of the cohort during the period, the average death rate, and the length of the period. But the average size of the cohort depends on how many of its members die, which in turn depends on the average size of the cohort. To straighten out this circular situation, a little algebra is needed. Let d be the average death rate times the length of the period under consideration, let c be the initial cohort size, let $\bar{c}$ be the average cohort size, and let c' be the final cohort size. Then

$$c' = c - d\bar{c} \tag{5}$$

and, to a reasonable approximation,

$$\bar{c} = \frac{c' + c}{2}. \tag{6}$$

Hence, we can write

$$\bar{c} = \frac{2c}{2 + d} \tag{7}$$

or

$$c' = c - \frac{2dc}{2 + d} \tag{8}$$
$$= \frac{(2 - d)}{(2 + d)} c,$$

---

53. You should figure out why this is an approximation and under what conditions it is a very good or a very poor one. See Exercise 5.

For our case, for the youngest age cohort,

$$d = 10\left[\frac{7.9 + 0.98}{2}\right]10^{-3} \tag{9}$$

$$= 0.044.$$

The factor of 10 arises because of the ten-year period from 1980 to 1989. The second factor, in brackets, is the average of the age-specific death rates for ages 0–9 and 10–19. The factor of 10^{-3} is there because the death rate is given in deaths-per-thousand and the population that d will multiply is in units of 10^6. The second entry in $\mathbf{P}(1990)$ will then be[54]

$$\frac{(2 - 0.044)}{(2 + 0.044)}(235) = 0.957\ (235) = 224.9. \tag{10}$$

Continuing in this way, you will obtain the result

$$\mathbf{P}(1990) = \begin{vmatrix} 101.5 \\ 224.9 \\ 221.1 \\ 178.4 \\ 119.5 \\ 87.8 \\ 57.0 \\ 26.6 \\ 7.6 \end{vmatrix}. \tag{11}$$

The sum of these numbers is 1,024.4, and so the total population in 1990 is about 1.024×10^9. Even though a one-child-per-family policy is implemented, the population has still grown. This is an example of a phenomenon called "population momentum." It occurs because of the age structure of the 1980 population, which is heavily weighted toward young people. The population will not start to decline until the age group with the low population reflecting the new family policy (ages 0–9 in 1990) grows into the childbearing years. The subsequent value of $\mathbf{P}$ can be obtained in the same way. The next five population vectors are

---

54. We should keep three significant figures in the expression $(2 - d)/(2 + d)$ and one significant figure after the decimal place in the entries in $\mathbf{P}(t)$ for now; at the very end, we can round off to a significance level commensurate with the input data. If the entries for $\mathbf{P}(1990)$ are rounded off too soon, then when $\mathbf{P}(2000)$ is calculated an error could result.

$$\mathbf{P}(2000) = \begin{vmatrix} 111.5 \\ 97.1 \\ 222.0 \\ 216.7 \\ 172.0 \\ 110.4 \\ 72.5 \\ 36.1 \\ 8.4 \end{vmatrix}, \mathbf{P}(2010) = \begin{vmatrix} 79.7 \\ 106.7 \\ 95.8 \\ 217.6 \\ 208.9 \\ 159.0 \\ 91.2 \\ 48.1 \\ 11.4 \end{vmatrix}, \mathbf{P}(2020) = \begin{vmatrix} 50.6 \\ 76.3 \\ 105.3 \\ 93.9 \\ 209.8 \\ 193.0 \\ 131.3 \\ 57.7 \\ 15.2 \end{vmatrix},$$

total = 1,046.7   1,018.4   933.0

(12)

$$\mathbf{P}(2030) = \begin{vmatrix} 45.4 \\ 48.4 \\ 75.3 \\ 103.2 \\ 90.5 \\ 193.9 \\ 159.4 \\ 83.1 \\ 18.2 \end{vmatrix}, \text{ and } \mathbf{P}(2040) = \begin{vmatrix} 30.9 \\ 43.4 \\ 47.8 \\ 73.8 \\ 99.5 \\ 83.6 \\ 160.2 \\ 100.9 \\ 26.3 \end{vmatrix}.$$

total = 817.4   666.4

Thus, sometime between the year 2030 and 2040, the goal of 7×10^8 will be obtained. Figure III-12 illustrates the progressive age-specific changes in the Chinese population.

The age structure of the new population is worth noting. In the year 2040, there will be more Chinese in the 60–69 and 70–79 age groups than in the 20–29 and 30–39 age groups. While this may seem

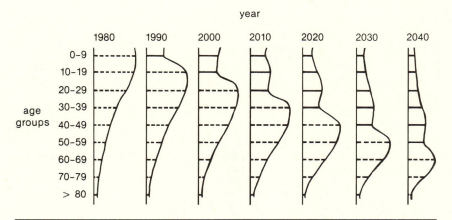

Figure III-12 The age distribution of the Chinese population for each decade from 1980 to 2040 under the assumptions stated in the text.

to pose a welfare problem, it should also be noted that there will be fewer youngsters to care for. In the year 1980, the ratio of the working population (assumed to be the population between the ages of 20 and 59) to the dependent population (assumed to be those under 20 and those 60 or over) was 0.88. In 2040, that ratio will be 0.84. Thus, if children and old folks are assumed to present a roughly similar welfare burden on society, the overall situation will not be so different in 2040 from what it is now.

If the one-child-per-family policy were to continue, the population would continue to shrink. At about the time the goal is reached, a two-child-per-family policy (or some close approximation) will presumably be put in effect. Exercise 6 asks you to look at the subsequent evolution of the population under a two-child-per-family policy. Even in this case the population continues to shrink. The reason is that some females will not live to childbearing age and thus will not replace themselves. See Exercise 7 for a hint on how to calculate the family size necessary to bring about exact replacement and a stationary population.[55]

The vectors $\mathbf{P}(t)$ can be calculated in a way slightly more sophisticated than the one shown above. Because we have expressed the population as a vector, we can construct a matrix, $\mathbf{M}$, which, if multiplied[56] by $\mathbf{P}(t)$, gives $\mathbf{P}(t + 10)$. That matrix[57] is

$$\mathbf{M} = \begin{vmatrix} 0 & 0.25 & 0.25 & 0 & 0 & 0 & 0 & 0 & 0 \\ 0.957 & 0 & 0 & 0 & 0 & 0 & 0 & 0 & 0 \\ 0 & 0.987 & 0 & 0 & 0 & 0 & 0 & 0 & 0 \\ 0 & 0 & 0.980 & 0 & 0 & 0 & 0 & 0 & 0 \\ 0 & 0 & 0 & 0.964 & 0 & 0 & 0 & 0 & 0 \\ 0 & 0 & 0 & 0 & 0.924 & 0 & 0 & 0 & 0 \\ 0 & 0 & 0 & 0 & 0 & 0.826 & 0 & 0 & 0 \\ 0 & 0 & 0 & 0 & 0 & 0 & 0.633 & 0 & 0 \\ 0 & 0 & 0 & 0 & 0 & 0 & 0 & 0.316 & 0 \end{vmatrix}. \quad (13)$$

EXERCISE 1: Derive Eq. 11.

EXERCISE 2: Explain why each entry in the matrix $\mathbf{M}$ is correct.

---

55. A stationary population is one in which the total population and the relative age distribution of the population are constant in time.
56. Rules for multiplying matrices by vectors are reviewed in Swartz (1973).
57. $\mathbf{M}$ is called a Leslie matrix in the demographic literature. A good reference for those interested in learning more about mathematical methods for modeling the age structure of populations is Keyfitz (1968). The same author (Keyfitz 1984) has also analyzed several of China's population options in a recent article.

EXERCISE 3: Explain what is wrong with the following residence time argument: The stock of population in 1980 is 10^9 and the outflow is 7.4 per thousand or 7.4×10^6; therefore the residence time of an individual in the population is $10^9/(7.4 \times 10^6)$, which equals 135 years.

EXERCISE 4: Explain why the number of deaths per decade among the people in a given cohort will not be exactly equal to $(2dc)/(2 + d)$ as calculated in Eq. 8. For which cohorts in 1980 will this be a good approximation and for which will it be a poor one?

* *EXERCISE 5:* If the age-specific death rates do not change, what are the odds that a Chinese born today will live to be at least (a) 60, (b) 70, and (c) 80 years old? Hint: If the probability of dying between ages of 0 and 9 is p, then $1 - p$ is the probability of surviving to reach age 10.

* *EXERCISE 6:* How would the Chinese population evolve if the one-child-per-family policy continues to the year 2040 and then a two-child-per-family policy is implemented? Calculate the answer by using the matrix approach, but modify the matrix in Eq. 13 to allow for two-child families (replacing the 0.25 at the top of the second and third columns by 0.5). Derive the population vectors for the years 2050 through 2100. Do you think the population will reach a steady state?

* *EXERCISE 7:* A stationary population can result only if an appropriate family size is selected. The size of that population and the time it takes to reach steady state from any given initial population will depend on the population size and age structure at the time the appropriate family-size policy is implemented, but the age distribution will depend only on the age-specific death rates. The appropriate family size necessary to get to a steady state also depends only on the age-specific death rates. What size family is necessary to get to a steady-state population? Solving this by trial and error is tedious. Instead, let the number at the top of the second and third columns of **M** be designated X. By writing and solving a set of algebraic equations expressing the fact that the populations in the 0–9, 10–19, and 20–29 age groups are in steady state, X can be determined. What number of children per family does that value of X correspond to? In the steady state, what will be the ratios of the populations in consecutive age groups?

** *EXERCISE 8:* Devise a family-size policy for China that will lead from the 1980 population to an eventual steady-state population of about 7×10^8 within 100 years.

15. An Unanswered Question

The linkages among the people and nations of the world resemble those within ecosystems. Symbiotic and competitive types of links exist among people within nations as they do among organisms in populations, and the symbiotic, competitive, and predator-prey links among nations in the world are analogous to certain relationships among populations in an ecosystem. Moreover, there are flows of information, energy, and materials within both systems and these flows are characterized by a multiplicity of time constants. There are both long loops in society (such as those involving feedback from the educational process) and short ones (such as the time you're given to pay most bills). Similarly, in ecosystems, some materials cycle quickly (phosphorus in a lake) and others more slowly (CO_2 through the biogeosphere). Both human societies and ecosystems contain provisions for storage of essential materials and information. We use banks, silos, dammed streams, and libraries, for example, while natural storage occurs in rich soils, aquifers, and the germ plasm of living organisms. Finally, there are both centralized and decentralized patterns of organization in human society and in ecosystems.

When complex systems are disturbed, the way they respond depends on how they are put together. What insights can you acquire into the consequences, both for ecosystems and for human society, of various patterns of organization? Before you address this question, delve into the literature on the stability of systems (start with May, 1973) to pick up some conceptual tools beyond those used in this book. In May's book you will learn more about how the eigenvalues discussed in Problem III.5 provide information about the stability of a system. You will also come across the notion of a "stability indicator." This is simply a property of a system that bestows stability on it. It was once widely believed that species diversity bestowed stability on ecosystems (i.e., diversity was a stability indicator). Now, theoretical ecologists are more inclined toward the view that the stability of an ecosystem is what allows a rich diversity of species to survive. As you delve further in the literature, you will also discover many different notions of what "stability" means. Which of them would you consider to be desirable attributes of human society?

Some interesting themes pertinent to the survival of human society that you may want to pursue are:

1. The consequences of the increasing number and strength of international linkages (international trade, multinational corporations, and international migration of workers).

2. The consequences of basing economies on stock resources (fossil fuels, nonrecycled raw materials) versus flow resources (renewable energy sources, such as solar; recycled materials).

3. The consequences of the shortening delay times between actions and reactions (feedbacks) in diplomacy, finance, and in news coverage.

4. The adequacy of buffering provided by presently available storage facilities for essential resources.

Appendix
Useful Numbers

Two types of data are listed below. Data of the first type are very accurately known, generally to better than ±1%. These include conversion factors relating different units of measurement (which are usually definitions and hence known perfectly); fundamental physical constants such as the Stefan-Boltzmann constant; and parameters characterizing physical properties of air, water, and radioactive isotopes. Data of the second type include rates of flows of energy and nutrients on Earth, concentrations of trace substance in various media, estimates of fossil fuel and biomass resources, and many others. These are known with less certainty—usually to no better than ±20%, and in some very important cases only to an order of magnitude. In cases where the degree of uncertainty is not likely to be obvious to the reader, explanatory notes are provided.

The data were culled from a wide variety of sources. Incorrect estimates and even typographical errors sometimes propagate through the literature. Thus two sources will occasionally differ about the value of a parameter by an amount greater than the experimental error. In such cases, I have exercised my best judgment about which value to use. A complete list of source material used for each subsection of the Appendix is provided at the end.

Appendix Contents

I. Units, Conversions, and Abbreviations

1. General prefixes

10	deka (da)	10^{-1}	deci (d)
10^2	hecto (h)	10^{-2}	centi (c)
10^3	kilo (k)	10^{-3}	milli (m)
10^6	mega (M)	10^{-6}	micro (μ)
10^9	giga (G)	10^{-9}	nano (n)
10^{12}	tera (T)	10^{-12}	pico (p)
10^{15}	peta (P)	10^{-15}	femto (f)
10^{18}	exa (E)	10^{-18}	atto (a)

2. Length

1 meter (m) = 100 centimeters (cm) = 3.281 feet (ft) = 39.37 inches (in)

1 mile = 5280 ft = 1.609 kilometers (km)

1 micron (μ) = 10^{-6} m

1 angstrom (Å) = 10^{-10} m

3. Area

1 hectare (ha) = 10^4 square meters (m^2) = 2.47 acres

1 acre = 43,560 square feet (ft^2)

1 barn (b) = 10^{-24} cm^2

4. Volume

1 cubic meter (m^3) = 1000 liters = 264.2 U.S. gallons = 35.31 cubic feet (ft^3)

1 liter (l) = 10^3 cubic centimeters (cm^3 or ml) = 1.057 U.S. quarts

1 acre foot = 1.234×10^3 m^3

1 cord = 128 ft^3

1 board foot = 2.36×10^{-3} m^3

1 cubic mile = 4.17 cubic kilometers (km^3)

1 barrel of petroleum (bbl) = 42 U.S. gallons = 0.159 m^3

5. Angles

360 degrees (°) = 2π radians

1 degree = 60 minutes (′) of arc

1 minute of arc = 60 seconds (″) of arc

6. Time

1 year (y or yr) = 3.1536×10^7 seconds (s or sec)

  = 8.76×10^3 hours (h or hr)

1 day (d) = 8.64×10^4 sec = 1440 minutes (min)

7. Mass

1 kilogram (kg) = 2.205 pounds (lb)

1 metric ton (tonne or MT) = 10^3 kilograms (kg)

    = 1.102 short tons

    = 0.9842 long tons

1 pound (lb) = 16 ounces avoirdupois (oz) = 453.6 grams (g)

8. Energy

1 joule (J) = 1 kg m^2/sec^2

    = 10^7 ergs = 0.2390 calories (cal)

    = 9.484×10^{-4} British thermal units (Btu)

    = 1 watt-second (Ws)

    = 6.242×10^{18} electron volts (eV)

    = 1 newton-meter (Nm)

1 kilowatt-hour (kWh) = 3.6×10^6 J

    = 3414 Btu

1 quad = 10^{15} Btu = 1.05×10^{18} J

1 Calorie = 1 kilocalorie (Kcal) = 10^3 cal

1 therm = 10^5 Btu

1 foot pound = 1.356 J

1 kiloton of TNT (KT) = 4.2×10^{12} J

9. Power

1 watt (W) = 1 joule/second

1 horsepower (hp) = 0.746 kilowatts (kW)

10. Force

1 newton (N) = 1 kg m/sec^2 = 10^5 dynes (dyn)

11. Pressure

1 pascal = 1 N/m^2 = 1 J/m^3

1 bar = 10^5 pascal = 0.9869 atmospheres (atm)

1 atmosphere (atm) = 76 cm of mercury

    = 14.7 lb/in^2

    = 760 torr

12. Viscosity

1 poise (p) = 1 dyn-sec/cm^2 = 0.1 kg/m sec

13. Temperature

degrees Celsius* (°C) = 5/9 [degrees Fahrenheit (°F) − 32]

degrees Fahrenheit (°F) = 1.8 degrees Celsius (°C) + 32

Kelvin or absolute temperature scale. Kelvins (K) = degrees Celsius + 273.15

———

*Sometimes designated Centigrade

14. Radiation units

1 becquerel (Bq) = 1 nuclear transformation/sec

1 curie (Ci) = 3.7×10^{10} transformations/sec

1 rad (rd) = an absorbed radiation dose of 100 ergs/g of absorbing material

1 gray (Gy) = 100 rd

1 roentgen (R) = an exposure to gamma or X radiation that produces 2.58×10^{-4} coulomb (C) of electric charge (counting either positive or negative but not both) per kg of dry air.

rem: a measure of "dose equivalent," is given by the dose in rads multiplied by the Quality Factor (QF):

rems = rads × QF

QF = 1 for gamma rays (photons) and beta rays (electrons and positrons)

= 10 for fission neutrons and protons

= 20 for alpha particles (nuclei of helium atoms)

1 sievert (Sv) = 100 rem

15. Mathematical symbols

=	equals	$\approx$ or $\cong$	equals approximately
$\neq$	not equal to	$\equiv$	identical to
$>$	is greater than	$<$	is less than
$\geq$	is greater than or equal to	$\leq$	is less than or equal to
		$<<$	is much less than
$>>$	is much greater than	$\sum_{i=1}^{n}$	sum over i from 1 to n
∞	infinity		
$\pm$	plus or minus (e.g., 11 ± 2 is the range of real numbers between 9 and 13)	$\prod_{i=1}^{n}$	product over i from 1 to n
		ΔX	a small change in X
$a:b$	the ratio of a to b	$\left.\dfrac{\partial f(x,y)}{\partial x}\right\|_{\substack{x=a \\ y=b}}$	the partial derivative of the function, f, with respect to x evaluated at $x=a$ and $y=b$
ppm	parts per million (10^6)		
ppb	parts per billion (10^9)		
ppm(v)	parts per million by volume		
$\propto$	proportional to		

e = base of natural logarithm = 2.718281828

II. Some Fundamental Constants of Physics and Chemistry

constants	values
Stefan-Boltzmann constant (σ)	$5.669 \times 10^{-8} \dfrac{J}{m^2\text{-}K^4\text{-sec}}$
Avogadro's number (A or N)	6.02×10^{23} molecules/mole*
Ideal Gas constant (R)	$8.310 \dfrac{J}{mole\ K}$
Boltzmann's constant (k)	1.38×10^{-23} J/K
Speed of light in vacuum (c)	2.9979×10^8 m/sec
Planck's constant (h)	6.626×10^{-34} J sec
gravitational constant (G)	$6.67 \times 10^{-11} \dfrac{N\ m^2}{kg^2}$
mass of electron (M_e)	9.110×10^{-28} g
mass of proton (M_p)	1.673×10^{-24} g
mass of neutron (M_n)	1.675×10^{-24} g
charge of electron (e)	1.60210×10^{-19} coulombs (C)

*Throughout this appendix and the text, "mole" refers to a gram-mole, i.e., M grams of a substance where M is the substance's molecular mass. One mole of any gas at STP occupies a volume of 22.4 liters and contains Avogadro's number of molecules.

III. Earth's Vital Statistics

parameter	value
mass of Earth	5.98×10^{24} kg
mass of atmosphere	5.14×10^{18} kg
mass of stratosphere	0.5×10^{18} kg
mass of oceans	1.4×10^{21} kg
mass of water in atmosphere	1.3×10^{16} kg
mass of surface fresh water	1.26×10^{17} kg
mass of living organisms (dry weight)	1.3×10^{15} kg
number of moles of dry air in atmosphere	1.8×10^{20}
average height in atmosphere at which pressure is one half of sea level pressure	5,600 m
average elevation of top of troposphere	12,000 m
mean oceanic depth	3,730 m
mean depth of oceanic mixed surface layer	75 m
mean elevation of continents	840 m
average distance between Earth and Sun	1.495×10^{11} m
equatorial radius	6.38×10^6 m
polar radius	6.36×10^6 m
total area	5.10×10^{14} m^2
area of continents	1.48×10^{14} m^2
Eurasia	0.536×10^{14} m^2
Africa	0.298×10^{14} m^2
North and Central America	0.238×10^{14} m^2
South America	0.179×10^{14} m^2

Antarctica	0.149×10^{14} m^2
Oceania	0.089×10^{14} m^2
ice-free land	1.33×10^{14} m^2
area of oceans	3.61×10^{14} m^2
ice-free Pacific Ocean	1.66×10^{14} m^2
ice-free Atlantic Ocean	0.83×10^{14} m^2
Indian Ocean	0.65×10^{14} m^2
ice-free Arctic ocean	0.14×10^{14} m^2
sea ice (average)	0.33×10^{14} m^2
volume of oceans	1.35×10^{18} m^3
volume of mixed ocean layer	2.7×10^{16} m^3
mean density of Earth	5,500 kg/m^3
surface seawater density (15°C)	1,026 kg/m^3
acceleration of gravity at Earth's surface	9.8 m/sec^2
mean surface air temperature	288 K

IV. Astronomical Data

unit	value
1 parsec	3.084×10^{16} m
1 light year (ly)	9.46×10^{15} m
1 astronomical unit (AU) (mean radius of Earth's orbit)	1.49×10^{11} m
number of nucleons in the universe	10^{80}
radius of universe	10^{26} m
radius of sun	6.96×10^8 m
mass of sun	1.99×10^{30} kg
mean distance from Earth to moon	3.84×10^8 m
radius of moon	1.74×10^6 m
mass of moon	7.34×10^{22} kg
period of lunar revolution about Earth	2.36×10^6 sec

V. Air

1. Physical constants for dry air at STP*

constant	value
average molecular weight	28.96
specific heat	
at constant pressure	1,004.2 J/kg °C
at constant volume	719.6 J/kg °C
density	1.293 kg/m^3
viscosity	1.72×10^{-4} poise
coefficient of heat conductivity	0.0209 W/m °C
speed of sound in air	331.4 m/sec

---

*Standard temperature and pressure, denoted STP, is a temperature of 0°C and a pressure of 1 atm.

2. Composition of Earth's dry atmosphere (1983)*

gas	fraction by number of moles	fraction by weight
Nitrogen (N_2)	0.7808	0.7549
Oxygen (O_2)	0.2095	0.2314
Argon (Ar)	0.0093	0.0128
Carbon Dioxide (CO_2)	340 ppm	516 ppm
Neon (Ne)	18 ppm	12 ppm
Helium (He)	5.2 ppm	0.7 ppm
Methane (CH_4)	1.5 ppm	0.8 ppm
Krypton (Kr)	1.1 ppm	3.2 ppm
Hydrogen (H_2)	0.5 ppm	0.03 ppm
Nitrous Oxide (N_2O)	0.3 ppm	0.45 ppm
Carbon Monoxide (CO)	0.1 ppm	0.1 ppm
Ozone (O_3)	0.01 ppm	0.015 ppm
Nitrogen Dioxide (NO_2)	0.2 ppb	0.3 ppb
Sulfur Dioxide (SO_2)	0.2 ppb	0.4 ppb
Hydrogen Sulfide (H_2S)	0.05 ppb	0.05 ppb
Nitric Oxide (NO)	0.05 ppb	0.05 ppb
Ammonia (NH_3)	< 0.05 ppb	< 0.03 ppb

*Concentrations less than 1 ppm(v) are uncertain to $\pm 50\%$; all others are believed to be known to better than $\pm 10\%$. The mean fraction, by weight, of water vapor and cloud water in Earth's atmosphere is about 0.0025.

3. Some atmospheric time constants (order of magnitude only)

typical tropospheric residence time of particles with > 20 micron diameter	< 1 day
tropospheric residence time of many reactive or very soluble gases (e.g., SO_2, H_2S, NO_2, NO)	1 day
time for gases to mix vertically in the troposphere	10 days
typical tropospheric residence time of particles of < 1 micron diameter	> 100 days
time for interhemispheric mixing of tropospheric gases	1 year
mixing time within the stratosphere	10 years
tropospheric residence time of CO_2	10 years
tropospheric residence time of nonreactive gases that exit to stratosphere	10 years

VI. Water

1. General properties

property	value
density at 0°C	999.87 kg/m^3
at 3.98°C	1,000.00 kg/m^3
at 15°C	999.13 kg/m^3
at 25°C	997.07 kg/m^3
molecular weight	18.015
latent heat of fusion at 0°C	3.33×10^5 J/kg or 79.6 cal/g
latent heat of vaporization	
at 100°C	2.258×10^6 J/kg or 539.6 cal/g
at 17°C	2.459×10^6 J/kg
at 0°C	2.499×10^6 J/kg
specific heat of liquid water	
at 15° C	4,184 J/kg °C or 1 cal/g °C
specific heat of water vapor at 100°C, constant pressure	2,008.3 J/kg °C or 0.48 cal/g °C
specific heat of ice	
at −2°C	2,100.4 J/kg °C or 0.502 cal/g °C
coefficient of heat conductivity	
at 100°C	0.683 W/m °C
at 17°C	0.595 W/m °C
at 0°C	0.563 W/m °C
viscosity	
at 100°C	2.8 millipoise
at 17°C	11.0 millipoise
at 0°C	17.5 millipoise

2. Stocks of water on Earth

stock	value (10^{15} m^3)
oceans	1,350
ice	29
groundwater*	8.3
freshwater lakes	0.125
saline lakes and inland seas	0.104
soil water	0.067
atmosphere	0.013

| water in living biomass | 0.003 |
| average amount in stream channels | 0.001 |

*About one half of the stock lies within a depth of 1 km.

3. Mean annual flows of water on Earth

flow	value (10^{12} m³/yr)
world precipitation on land	108
world precipitation on the sea	410
world evaporation from the sea	456
world evapotranspiration from the land	62
world runoff	46
U.S. precipitation*	5.6
U.S. evapotranspiration*	3.95
U.S. runoff*	1.65

*Excluding Alaska and Hawaii

4. Water used by human beings

use	withdrawal (10^9 m³/yr)*		consumption (10^9 m³/yr)*	
	world	*U.S.†*	*world*	*U.S.†*
municipal and domestic	220	42	75	10
mining and manufacturing	390	52	40	6
electric power plant cooling	620	180	8	2.6
irrigation and live stock	2,100	205	1,100	125

*Water consumed is water rendered unavailable for direct further use; water withdrawn is water taken from a water supply but not necessarily consumed. Values here are estimated for 1980 and are uncertain to ± 10% for the United States, and ± 25% for the world.
†Excluding Alaska and Hawaii

VII. Energy

1. Energy flows*

flow	value (10^{12} W)
energy radiated by sun into space	3.7×10^{14}
solar radiation incident on the top of Earth's atmosphere	175,000
solar radiation reflected back to space from Earth	53,000
solar radiation reflected back to space from Earth's atmosphere	46,000

solar radiation absorbed in atmosphere (about 80% of this is absorbed in air and dust, and about 20% in cloud water)	44,000
rate at which latent heat flows from Earth's surface to atmosphere	42,000
rate at which infrared radiation leaving Earth's surface flows directly to space	10,200
rate at which convective heat flows from Earth's surface to atmosphere	8,600
wind, waves, ocean currents	500–2,000
net primary productivity on Earth	75–125
energy conducted from Earth's interior to its surface	20–40
world energy consumption (1980)	10
U.S. energy consumption (1980)	2.5
energy content of food consumed by world's human population (1980)	0.55
world electricity production (1980)	0.87
U.S. electricity production (1980)	0.26

*Values are uncertain to roughly $\pm$ 5% or less, except as indicated.

2. Earth's nonrenewable energy resources

resource	estimated stock (1980) (10^{21} J)	consumption (1980) (10^{18} J/yr)*	
		world	*U.S.*
petroleum	10	135	41
natural gas	10	60	20
coal	250	90	15
tar sands	>2	0	0
oil shale	2,000	0	0
uranium in non-breeding light water reactors	20	6.3	3.1
thorium and uranium in breeder reactors	10,000	0	0
deuterium and lithium in seawater (for fusion power)	10^{10}	0	0

*When fuels are used for electricity generation, the heat energy rather than the electrial energy is quoted. When the world figures are summed, the total is less than the value given in Table VI.1 for world energy consumption because Table VI.2 does not include use of renewables. Worldwide combustion of fuel wood and dung in 1980 produced about 30×10^{18} J and hydroelectric power produced about 6.1×10^{18} J of electricity. In the United States, hydropower yielded about 1.0×10^{18} J of electricity. Stocks are uncertain to $\pm$ 50%; consumption figures are reliable to ± 1% in the United States and ± 10% worldwide.

3. Average composition of fossil fuels

fuel	constituent	percentage*
coal†	$CH_{0.8}$	75
	H_2O	13
	ash	9
	S	2.5
	N	1.0
	Al	0.5
	Ca	0.5
	Mn	0.01
	Zn	0.005
	Pb	0.001
	Ni	0.001
	Cr	0.001
	Cu	0.001
	As	0.001
	Mo	0.0005
	Se	0.0001
	U	0.0001
	Hg	0.00001
	Cd	0.00001
petroleum (crude)	$CH_{1.5}$	98
	S	1.5
	N, O_2	<0.5
	Ni	0.001
	Mo	0.001
	Ca	0.001
natural gas	CH_4	75
	C_2H_6	6
	C_3H_8	4
	C_4H_{10}	2
	C_5H_{12}	1
	noncombustibles	12

---

*Percentages are by weight for coal and petroleum, and by number of moles for natural gas.

†Percentages add up to greater than 100 because the ash fraction includes some of the trace quantities listed below ash.

4. Energy content of selected substances

substance	energy content (10^6 J/kg)*
natural gas	3.9×10^7 J/m^3 (STP)
gasoline	48
petroleum (crude)	43
	(6.1×10^9 J/bbl)
typical animal fat	38
coal	29.3
charcoal	29
paper	20
dry biomass	16
air-dried wood or dung	15
crop wastes (20% moisture)	13
bread	12
milk	3.0
beer	1.8

*Except where noted

VIII. The Elements

1. Abundance of the elements

element	Earth's crust (ppm)*	seawater (ppm)*	biomass (ppm)*†
O	456,000	857,000	630,000
Si	273,000	3	15,000
Al	83,600	0.01	500
Fe	62,200	0.01	1,000
Ca	46,600	400	40,000
Mg	27,640	1,350	4,000
Na	22,700	10,500	2,000
K	18,400	380	20,000
Ti	6,320	0.001	100
H	1,520	108,000	80,000
P	1,120	0.07	5,000
Mn	1,060	0.0002	100
F	554	1.3	50
Ba	390	0.03	300
Sr	384	8.1	200
S	340	885	5,000
C	180	28	200,000
V	136	0.002	—
Cl	131	19,000	2,000
Cr	122	0.00005	—
Ni	99	0.005	5
Zn	76	0.01	5
Cu	68	0.003	20
N	19	0.50	20,000
Pb	13	0.000005	< 0.2
B	9	4.6	100
Br	2.5	65	10
U	2.3	0.003	—
As	1.8	0.003	3
Hg	0.09	0.00003	—

* Parts per million on a weight-per-weight basis

† Based on assumption of 50% water content of global biomass stock

2. Physical properties of the elements

name	symbol	atomic number	average atomic weight	atomic* weight of dominant isotopes	boiling† point (°C)	melting† point (°C)	density‡ 10^3kg/m^3	specific heat J/(kg) (K)
Actinium	Ac	89	227	227	—	1,050	—	—
Aluminum	Al	13	26.98	27	2,450	660	2.70	899
Americium	Am	95	243	**	—	—	11.7	138
Antimony	Sb	51	121.8	121, 123	1,380	630.5	6.62	205
Argon	Ar	18	39.95	40, 36, 38	−185.8	−189.2	1.40	523
Arsenic	As	33	74.92	75	613	817	5.72	343
Astatine	At	85	210	219	—	(302)	—	—
Barium	Ba	56	137.3	138, 137, 136, 135, 134, 130, 132	1,640	714	3.5	284
Berkelium	Bk	97	247	**	—	—	—	—
Beryllium	Be	4	9.01	9	2,770	1,277	1.85	1,881
Bismuth	Bi	83	209	209	1,560	271.3	9.8	142
Boron	B	5	10.81	11, 10	—	(2,030)	2.34	1,292
Bromine	Br	35	79.90	79, 81	58	−7.2	3.12	293
Cadmium	Cd	48	112.4	114, 112, 111, 110, 113, 116, 106, 108	765	320.9	8.65	230

Element	Symbol	Atomic Number	Atomic Weight	Isotopes				
Calcium	Ca	20	40.08	40, 44, 42, 48, 43	1,440	838	1.55	623
Californium	Cf	98	**249**	**	—	—	—	—
Carbon	C	6	12.01	12, 13, **14**	48.3	3,727	2.26	690
Cerium	Ce	58	140.1	140, 142, 138, 136	3,468	795	6.67	176
Cesium	Cs	55	132.9	133	690	28.7	1.90	217
Chlorine	Cl	17	35.45	35, 37	−34.7	−101.0	1.56	485
Chromium	Cr	24	52.00	52, 53, 50, 54	2,665	1,875	7.19	460
Cobalt	Co	27	58.93	59	2,900	1,495	8.9	414
Copper	Cu	29	63.55	63, 65	2,595	1,083	8.96	385
Curium	Cm	96	**247**	**	—	—	—	—
Dysprosium	Dy	66	162.5	164, 162, 163, 161, 160, 158, 156	2,600	1,407	8.54	171

*This column includes the weights of naturally occurring isotopes. Stable isotopes that occur at levels of abundance greater than 0.01% are listed in order of decreasing abundance. Naturally occurring radioactive isotopes are listed in boldface and they follow the stable isotopes except in those cases where they occur at levels comparable to those of the stable isotopes. For properties of some of the more important artificial, radioactive isotopes, see Section XI, Table 1, of this Appendix.

**Indicates a synthetically produced element with no naturally occurring isotopes.

†Values are for 1 atmosphere pressure. Parentheses indicate that the value is for the most stable or best known isotope.

‡For elements that are solid or liquid at standard temperature and pressure (STP), these values reflect density at STP. Where elements occur in the gas phase at STP, the value listed is the density of that element's liquid phase at the boiling point under 1 atmosphere pressure.

name	symbol	atomic number	average atomic weight	atomic* weight of dominant isotopes	boiling† point (°C)	melting† point (°C)	density† 10³kg/m³	specific heat (J/kg)
Einsteinium	Es	99	**254**	**	—	—	—	—
Erbium	Er	68	167.3	166, 168, 167, 170, 164, 162	2,900	1,497	9.05	167
Europium	Eu	63	152.0	153, 151	1,439	826	5.26	163
Fermium	Fm	100	**257**	**	—	—	—	—
Fluorine	F	9	19.00	19	−188.2	−219.6	1.51	752
Francium	Fr	87	223	**223**	—	(27)	—	—
Gadolinium	Gd	64	157.3	158, 160, 156, 157, 155, 154, 152	3,000	1,312	7.89	297
Gallium	Ga	31	69.72	69, 71	2,237	29.8	5.91	330
Germanium	Ge	32	72.59	74, 72, 70, 73, 76	2,830	937.4	5.32	305
Gold	Au	79	197.0	197	2,970	1,063	19.3	130
Hafnium	Hf	72	178.5	180, 178, 177, 179, 176, 174	5,400	2,222	13.1	146
Hahnium	Ha	105	**260**	**	—	—	—	—
Helium	He	2	4.00	4	−268.9	−269.7	0.12	5,225
Holmium	Ho	67	164.9	165	2,600	1,461	8.80	163

Element	Symbol							
Hydrogen	H	1	1.01	1, 2, 3	−252.7	−259.2	0.07	14,421
Indium	In	49	114.8	**115**, 113	2,000	156.2	7.31	238
Iodine	I	53	126.9	127	183	113.7	4.94	217
Iridium	Ir	77	192.2	193, 191	5,300	2,454	22.5	130
Iron	Fe	26	55.85	56, 54, 57, 58	3,000	1,536	7.86	460
Krypton	Kr	36	83.80	84, 86, 83, 82, 80, 78	−152	−157.3	2.6	—
Kurchatorium	Ku	104	**260**	**	—	—	—	—
Lanthanum	La	57	138.9	139, **138**	3,470	920	6.17	188
Lawrencium	Lr	103	**257**	**	—	—	—	—
Lead	Pb	82	207.2	208, 206, 207, 204, **210**	1,725	327.4	1.4	130
Lithium	Li	3	6.94	7, 6	1,330	180.5	0.53	3,302
Lutetium	Lu	71	175.0	175, **176**	3,327	1,652	9.84	155
Magnesium	Mg	12	24.31	24, 26, 25	1,107	650	1.74	1,045
Manganese	Mn	25	54.94	55	2,150	1,245	7.43	481
Mendelevium	Md	101	**256**	**	—	—	—	—
Mercury	Hg	80	200.6	202, 200, 199, 201, 198, 204, 196	357	−38.4	13.6	138

name	symbol	atomic number	average atomic weight	atomic* weight of dominant isotopes	boiling† point (°C)	melting† point (°C)	density† 10^3kg/m³	specific heat (J/kg)
Molybdenum	Mo	42	95.94	98, 96, 95, 92, 100, 97, 94	5,560	2,610	10.2	255
Neodymium	Nd	60	144.2	142, 144, 146, 143, 145, 148, 150, 147	3,027	1,024	7.00	188
Neon	Ne	10	20.18	20, 22, 21	−246	−248.6	1.20	—
Neptunium	Np	93	**237**	**	—	637	19.5	—
Nickel	Ni	28	58.71	58, 60, 62, 61, 64	2,730	1,453	8.9	439
Niobium	Nb	41	92.91	93	3,300	2,468	8.4	272
Nitrogen	N	7	14.01	14, 15	−195.8	−210	0.81	1,033
Nobelium	No	102	**254**	**	—	—	—	—
Osmium	Os	76	190.2	192, 190, 189, 188, 187, 186, 184	5,500	3,000	22.6	130
Oxygen	O	8	16.00	16, 18, 17	−183	−218.8	1.14	911

Element	Symbol							
Palladium	Pd	46	106.4	106, 108, 105, 110, 104, 102	3,980	1,552	12.0	242
Phosphorus	P	15	30.97	31	280	44.2	1.82	740
Platinum	Pt	78	195.09	195, 194, 196, 198, **190**	4,530	1,769	21.4	134
Plutonium	Pu	94	**244**	**	3,235	640	—	—
Polonium	Po	84	**209**	**210**	—	254	9.2	—
Potassium	K	19	39.10	39, 41, **40**	760	63.7	0.86	740
Praseodymium	Pr	59	140.9	141	3,127	935	6.77	201
Promethium	Pm	61	**145**	**	—	(1,027)	—	—
Protactinium	Pa	91	**231**	**231**	—	(1,230)	15.4	—
Radium	Ra	88	**226**	**226**	—	700	5.0	—
Radon	Rn	86	**222**	**222**	(−61.8)	(−71)	—	—
Rhenium	Re	75	186.2	**187**, 185	5,900	3,180	21.0	138
Rhodium	Rh	45	102.9	103	4,500	1,966	12.4	247
Rubidium	Rb	37	85.46	85, **87**	688	38.9	1.53	334
Ruthenium	Ru	44	101.1	102, 104, 101, 100, 99, 96, 98	4,900	2,500	12.2	238
Samarium	Sm	62	150.4	152, 154, **147**, 147, 148, 150, 144	1,900	1,072	7.54	176

name	symbol	atomic number	average atomic weight	atomic* weight of dominant isotopes	boiling† point (°C)	melting† point (°C)	density† 10³kg/m³	specific heat (J/kg)
Scandium	Sc	21	44.96	45	2,730	1,539	3.0	543
Selenium	Se	34	78.96	80, 78, 76, 82, 77, 74	685	217	4.79	351
Silicon	Si	14	26.09	28, 29, 30	2,680	1,410	2.33	677
Silver	Ag	47	107.9	107, 109	2,210	960.8	10.5	234
Sodium	Na	11	22.99	23	892	97.8	0.97	1,233
Strontium	Sr	38	87.62	88, 86, 87, 84	1,380	768	2.6	736
Sulfur	S	16	32.06	32, 34, 33, 36	444.6	119.0	2.07	732
Tantalum	Ta	73	180.9	181, 180	5,425	2,996	16.6	151
Technetium	Tc	43	**96.91**	**	—	2,140	11.5	—
Tellurium	Te	52	127.6	130, 128, 126, 125, 124, 122, 123, 120	989.8	449.5	6.24	197
Terbium	Tb	65	158.9	159	2,800	1,356	8.27	184
Thallium	Tl	81	204.4	205, 203,	1,457	303	11.85	130
Thorium	Th	90	**232**	**232, 228**	3,850	1,750	11.7	142
Thulium	Tm	69	168.9	169	1,727	1,545	9.33	159

Name	Symbol			Isotopes				
Tin	Sn	50	118.7	120, 118, 116, 119, 117, 124, 122, 112, 114, 115	2,270	231.9	7.3	226
Titanium	Ti	22	47.90	48, 49, 50, 46, 47	3,260	1,668	4.51	527
Uranium	U	92	**238**	**238, 235**	3,818	1,132	19.07	117
Vanadium	V	23	50.94	51, 50	3,450	1,900	6.1	502
Wolfram	W	74	183.9	184, 186, 182, 183, **180**	5,930	3,410	19.3	134
Xenon	Xe	54	131.3	132, 129, 131, 134, 136, 130, 128, 124, 126	− 108.0	− 111.9	3.06	—
Ytterbium	Yb	70	173.0	174, 172, 173, 171, 176, 170, 168	1,427	824	6.98	146
Yttrium	Y	39	88.91	89	2,927	1,509	4.47	297
Zinc	Zn	30	65.38	64, 66, 68, 67, 70	906	419.5	7.14	383
Zirconium	Zr	40	91.22	90, 94, 92, 91, 96	3,580	1,852	6.49	276

IX. Global Natural Background Flow to the Atmosphere of Selected Substances*

substance	rate (kg of substance/yr)†
CH_4	7×10^{11}
H_2S and SO_2	10^{11} kg(S)/yr
SO_4^{-2}	5×10^{10} kg(S)/yr
NO_x and NH_3	5×10^{11} kg(N)/yr
particles less than 20 microns in diameter	3×10^{12}
arsenic	2×10^7
cadmium	3×10^5
chromium	6×10^7
copper	2×10^7
lead	6×10^6
manganese	6×10^8
mercury	3×10^7
nickel	3×10^7
selenium	3×10^6
vanadium	7×10^7
zinc	4×10^7

*This includes dust emissions, volcanic eruptions, biological processes, and volatilization from land and water. The first three entries in the table are believed to be known to ± 30%. The others are far more uncertain and are order-of-magnitude estimates only.
†Except where noted

X. Chemical Reactions and Constants

1. Some important chemical reactions*

$106\ CO_2 + 16\ NO_3^- + HPO_4^{-2} + 18\ H^+ + 122\ H_2O$ photosynthesis,
$\rightleftharpoons C_{106}H_{263}O_{110}N_{16}P_1 + 138\ O_2$ respiration

$2\ N_2 + 6\ H_2O \rightarrow 4\ NH_3 + 3\ O_2$ nitrogen fixation

$NH_4^+ + 2\ O_2 \rightarrow NO_3^- + 2\ H^+ + H_2O$ nitrification

$4\ HNO_3 + 5\ CH_2O \rightarrow 5\ CO_2 + 7\ H_2O + 2\ N_2$ denitrification

$H_2SO_4 + 2\ CH_2O \rightarrow H_2S + 2\ H_2O + 2\ CO_2$ bacterial reduction of sulfate

$4\ FeS + 6\ H_2O + 11\ O_2 \rightarrow 4\ Fe(OH)_3 + 4\ H_2SO_4$ ⎤ acid production
$4\ FeS_2 + 8\ H_2O + 15\ O_2 \rightarrow 2\ Fe_2O_3 + 8\ H_2SO_4$ ⎦ from ores (acid mine drainage)

$$NO_2 + light \rightarrow NO + O$$
$$O + O_2 \rightarrow O_3$$

ozone formation (in presence of hydrocarbon radicals, this reaction can lead to a significant increase in tropospheric ozone)

$$NO + O_3 \rightarrow NO_2 + O_2$$
$$NO_2 + O \rightarrow NO + O_2$$

catalytic cycle by which NO_x reduces stratospheric ozone

$$6\ NO + 4\ NH_3 \rightarrow 6\ H_2O + 5\ N_2$$

scrubbing of NO with ammonia

$$H_2O + SO_2 + CaCO_3 \rightarrow CaSO_4 + H_2CO_3$$

scrubbing of SO_2 with limestone

$$SiO_2 + 2\ H_2O \rightarrow H_4SiO_4$$

dissolution of quartz

$$CaCO_3 + H_2O \rightarrow Ca^{+2} + HCO_3^- + OH^-$$

dissolution of calcite

$$CaCO_3 + H_2CO_3 \rightarrow Ca^{+2} + 2\ HCO_3^-$$

carbonic acid weathering of calcite

$$2\ KAlSi_3O_8 + 2\ H_2CO_3 + H_2O$$
$$\rightarrow Al_2Si_2O_5(OH)_4 + 4\ SiO_2 + 2\ K^+ + 2\ HCO_3^-$$

carbonic acid weathering of feldspar to kaolinite

$$3\ Ca_{0.33}Al_{4.67}Si_{7.33}O_{20}(OH)_4 + 2\ H^+ + 23\ H_2O$$
$$\rightarrow 7\ Al_2Si_2O_5(OH)_4 + 8\ H_4SiO_4 + Ca^{+2}$$

acid weathering of calcium montmorillonite to kaolinite

$$Al_2Si_2O_5(OH)_4 + 5\ H_2O$$
$$\rightarrow 2\ H_4SiO_4 + Al_2O_3 \cdot 3\ H_2O$$

dissolution of kaolinite to gibbsite

$$Al_2O_3 \cdot 3\ H_2O + 2\ H_2O \rightarrow 2\ Al(OH)_4^- + 2\ H^+$$

dissolution of gibbsite

---

*The first six of these reactions are carried out biologically. The actual processes are more complex than is indicated by the expressions above, which represent the net reaction only. For example, nitrification is carried out in two steps: the bacteria, *Nitrosomonas*, converts NH_4^+ to NO_2^-, and *Nitrobacter* converts NO_2^- to NO_3^-.

2. Some chemical equilibrium dissociation constants (at 25°C)*

reaction	10^{-pK}
$H_2O \rightleftharpoons H^+ + OH^-$	10^{-14}
$H_2CO_3 \rightleftharpoons H^+ + HCO_3^-$	$10^{-6.35}$
$HCO_3^- \rightleftharpoons H^+ + CO_3^{-2}$	$10^{-10.33}$
$HCl \rightleftharpoons H^+ + Cl^-$	$10^{3.0}$
$H_2SO_4 \rightleftharpoons H^+ + HSO_4^-$	$10^{3.0}$
$HNO_3 \rightleftharpoons H^+ + NO_3^-$	$10^{1.0}$
$HSO_4^- \rightleftharpoons H^+ + SO_4^-$	$10^{-1.9}$
$H_2SO_3 \rightleftharpoons H^+ + HSO_3^-$	$10^{-1.77}$
$HSO_3^- \rightleftharpoons H^+ + SO_3^{-2}$	$10^{-7.21}$
$NH_3 + H_2O \rightleftharpoons NH_4^+ + OH^-$	$10^{-4.74}$
$H_3BO_3 \rightleftharpoons H^+ + H_2BO_3^-$	$10^{-9.3}$

*See the introduction to Chapter II, Section C for a discussion of how these are used in chemical equilibrium calculations.

3. Some values of Henry's constant* (at 25°C)

equilibrium ratio	Henry's constant (moles/liter-atm)
$\dfrac{[H_2SO_3]}{p(SO_2)}$	$10^{0.096}$
$\dfrac{[H_2CO_3]}{p(CO_2)}$	$10^{-1.47}$
$\dfrac{[H_2NO_3]}{p(NO_2)}$	$10^{-1.6}$
$\dfrac{[NH_3]}{p(NH_3)}$	$10^{1.76}$
$\dfrac{[CO]}{p(CO)}$	$10^{-3.0}$
$\dfrac{[N_2O]}{p(N_2O)}$	$10^{-1.59}$
$\dfrac{[H_2S]}{p(H_2S)}$	$10^{-0.97}$

*See the introduction to Chapter II, Section C for a discussion of how these are used in chemical equilibrium calculations.

4. Solubility products for solids* (at 25°C)

solid	solubility product (moles²/liter²)†	
calcite	$[Ca^{+2}][CO_3^{-2}] = 10^{-8.42}$	(fresh water)
aragonite	$[Ca^{+2}][CO_3^{-2}] = 10^{-8.22}$	(fresh water)
	$= 10^{-6.05}$	(seawater)
gypsum	$[Ca^{+2}][SO_4^{-2}] = 10^{-4.6}$	(fresh water)
dolomite	$[Ca^{+2}][Mg^{+2}][CO_3^{-2}]^2 = 10^{-16.7}$	(fresh water)

*See the introduction to Chapter II, Section C for a discussion of how these are used in chemical equilibrium calculations.
†The solubility product for dolomite has units of (moles⁴/liter⁴).

5. Equilibrium constants for acid-induced metal mobilization*

reaction	constant (liters/mole)†
$2H^+ + CuO \rightleftharpoons Cu^{+2} + H_2O$	$\dfrac{[Cu^{+2}]}{[H^+]^2} = 10^{7.7}$
$2H^+ + ZnO \rightleftharpoons Zn^{+2} + H_2O$	$\dfrac{[Zn^{+2}]}{[H^+]^2} = 10^{11.1}$
$2H^+ + Fe(OH)_2 \rightleftharpoons Fe^{+2} + 2H_2O$	$\dfrac{[Fe^{+2}]}{[H^+]^2} = 10^{12.9}$
$2H^+ + Cd(OH)_2 \rightleftharpoons Cd^{+2} + 2H_2O$	$\dfrac{[Cd^{+2}]}{[H^+]^2} = 10^{13.5}$
$2H^+ + Mn(OH)_2 \rightleftharpoons Mn^{+2} + 2H_2O$	$\dfrac{[Mn^{+2}]}{[H^+]^2} = 10^{15.2}$
$2H^+ + Mg(OH)_2 \rightleftharpoons Mg^{+2} + 2H_2O$	$\dfrac{[Mg^{+2}]}{[H^+]^2} = 10^{16.8}$
$2H^+ + Ca(OH)_2 \rightleftharpoons Ca^{+2} + 2H_2O$	$\dfrac{[Ca^{+2}]}{[H^+]^2} = 10^{22.8}$
$3H^+ + Al(OH)_3 \rightleftharpoons Al^{+3} + 3H_2O$	$\dfrac{[Al^{+3}]}{[H^+]^3} = 10^{8.5}$

*See Problem III.2 for a discussion of how these constants are used to estimate the effects of acidification on dissolved metal concentrations.
†The last reaction (aluminum) has an equilibrium constant in units of liters²/mole².

XI. Radiation and Radioactivity

1. Some important radioactive decay processes*

isotope	half-life	decay product	emitted radiation	maximum energy of radiation (MeV)
I^{131}	8.1 days	Xe^{131}	β^-	0.81
			γ	0.72
Sr^{89}	52 days	Y^{89}	β^-	1.5
Sr^{90}	28 yr	Y^{90}	β^-	0.59
Cs^{137}	30 yr	Ba^{137}	β^-	1.18
			γ	0.66
Pu^{239}	24,400 yr	U^{235}	α	5.15
			γ	0.05
U^{238}	4.5×10^9 yr	Th^{234}	α	4.18
			γ	0.05
C^{14}	5,700 yr	N^{14}	β^-	0.16
H^3 (tritium)	12.3 yr	He^3	β^-	0.018
Rn^{222}	3.8 days	Po^{218}	α	5.5
Po^{218}	3 min	Pb^{214}	α	6.0
Pb^{214}	27 min	Bi^{214}	β^-	0.7
			γ	0.35
Bi^{214}	20 min	Po^{214}	β^-	3.17
Po^{214}	0.00016 sec	Pb^{210}	α	7.68
Pb^{210}	22 yr	Bi^{210}	β^-	0.02
			γ	0.047

*The first four isotopes listed are among the major radioactive substances present in reactor cores or in radioactive wastes removed from these nuclear reactors; U^{238} is the dominant natural isotope of uranium; C^{14} is an important tracer in the environmental sciences and in archeology; and tritium is an important part of the radioactive debris from a hydrogen bomb explosion. The last six isotopes, which form a chain, are the critical reactions needed to estimate indoor air exposure from radioactive radon.

2. Human radiation exposure in the United States

natural sources	whole body annual dose (millirem/yr)
cosmic radiation*	30
C^{14}	1
terrestrial radionuclides†	
external to body	30
internal isotopes except radon daughters (mostly K^{40})	20
radon daughters in lungs††	100
anthropogenic sources	
diagnostic X rays	
dental	3
medical	95

therapeutic radiation	14
fallout from past atmospheric testing of nuclear weapons (dose in 1982)	2
television receivers	0.5
airline travel	0.5
nuclear energy (1980, worldwide general public only)	0.01

*This is the altitude-weighted average for the U.S. population.
†These figures vary geographically by a factor of at least two.
††Lung dose may be considerably higher for smokers.

XII. The Biosphere

1. Global biomass and productivity

location	living biomass stocks [10^{12} kg(C)]	dead organic matter [10^{12} kg(C)]	net primary productivity [10^{12} kg(C)/yr]
continental	560 + 300 − 100	1,500 ± 1,000	50 ± 15
oceanic	2 ± 1	2,000 ± 1,000	25 ± 10

2. Area, biomass, and productivity of ecosystem types

ecosystem type*	area (10^{12} m^2)	mean plant biomass [kg(C)/m^2]	average net primary productivity [kg(C)/m^2/yr]
tropical forests	24.5	18.8	0.83
temperate forests	12.0	14.6	0.56
boreal forests	12.0	9.0	0.36
woodland and shrubland	8.0	2.7	0.27
savanna	15.0	1.8	0.32
grassland	9.0	0.7	0.23
tundra and alpine meadow	8.0	0.3	0.065
desert scrub	18.0	0.3	0.032
rock, ice, and sand	24.0	0.01	0.015
cultivated land	14.0	0.5	0.29
swamp and marsh	2.0	6.8	1.13
lake and stream	2.5	0.01	0.23
open ocean	332.0	0.0014	0.057
upwelling zones	0.4	0.01	0.23
continental shelf	26.6	0.005	0.16
algal bed and reef	0.6	0.9	0.90
estuaries	1.4	0.45	0.81

*For a description of each of the major types of ecosystems (deserts, boreal forests, estuaries, etc.), see Whittaker (1970), Whittaker and Likens (1973), and Ehrlich et al. (1977).

XIII. Flows and Stocks of Carbon, Nitrogen, Phosphorus, and Sulfur on Earth*

1. Carbon flow

flow	magnitude 10^{12} kg(C)/yr
CO_2 flux to the atmosphere from decomposition and combustion of terrestrial organic matter and from animal respiration [This flow is nearly exactly balanced by a flow of inorganic carbon from the atmosphere to terrestrial living biomass in net primary productivity.]	50
inorganic carbon production in the oceanic mixed layer from decomposition of oceanic organic matter and animal respiration [This and the subsequent flow are nearly exactly balanced by a flow of inorganic carbon from seawater to living organisms in oceanic net primary productivity.]	20
inorganic carbon production in the deep ocean from decomposition of oceanic organic matter	5.0
net upwelling of inorganic carbon from deep ocean to the mixed oceanic layer	5.0
CO_2 flux to the atmosphere from fossil fuel burning and cement manufacturing	5.3
river flow of organic carbon to the oceans	0.2
deposition of carbon to oceanic sediment from sinking oceanic detritus	0.1
inorganic carbon production from weathering of rock and sediment	0.1

*Anthropogenic flows are believed to be known to within ±15%. Natural flows are often only crudely known. Most are uncertain to ±50%; and some, like the global biological nitrogen fixation rate, could be wrong by a factor of three (i.e., a value of 3 has a range of uncertainty from 1 to 9). Stocks in organic matter and in soil, rock, fuel, and sediments are believed to be known to within a factor of two. Atmospheric CO_2 and N_2 stocks are known to better than ±1%.

2. Nitrogen flow

flow	magnitude [10^{12} kg(N)/yr]
ammonification (production of NH_4^+ from organic nitrogen, the end stage of decomposition)	5
assimilation (conversion of NH_4^+ and NO_3^- to protein by vegetation and microbes; very roughly, one half of the nitrogen is assimilated as NH_4^+ and one half as NO_3^-, which was nitrified from NH_4^+)	5

natural background flow of NH_3 and NO_x from soil and water to the atmosphere	0.5
precipitation of NH_4^+ and NO_x to Earth's surface in rain and snow	0.1
denitrification (conversion of soil and water NO_3^- to atmospheric N_2 or N_2O, whose production rates are very roughly equal)	0.1
biological nitrogen fixation (about two thirds by continental organisms, one third by marine)	0.2
global anthropogenic nitrogen fixation in 1980 [Contributions from fossil fuel combustion was about one third of total; the remainder is mostly fertilizer production.]	0.1
river flow of fixed nitrogen to sea	0.01
fixation of atmospheric N_2 by lightning	0.01
production of stratospheric NO from N_2O	0.001

3. Sulfur flow

flow	magnitude [10^{12} kg(S)/yr]
plant uptake of sulfur	
continents	0.15
oceans	0.6
flow of SO_2 and SO_4^{-2} to Earth's surface, mostly in precipitation as H_2SO_4 and by dry deposition of aerosols	0.24
flow of S to the atmosphere from biological sources, sea spray, and volcanoes [about 33% of this flow is SO_4^{-2}, the remainder being H_2S and SO_2; the volcanic contribution is about 20% and the sea spray contribution is 25%.]	0.15
emissions to the atmosphere of SO_2 from fossil fuel burning, and metal-ore smelting (1980)	0.085
river flow of sulfur to the sea	0.1
fertilizer and industrial SO_4^{-2} production (1980)	0.03

4. Phosphorus flow

flow	magnitude [10^{12} kg(P)/yr]
uptake of PO_4^{-2} by living organisms (balanced by nearly equal rates of loss of phosphorus from living organisms by excretion and death)	
continental	0.2
marine	1.0

extraction of PO_4^{-2} from sediment for fertilizer, detergents, etc. (1980)	0.02
phosphorus in river flow to the sea	0.02
rate of guano deposition on land	0.0004
rate of extraction of phosphorus from the sea in harvested fish (1980)	0.0004

5. Major stocks of carbon

stock	magnitude [10^{12} kg(C)]
carbon in rock and sediment	10^7
carbon in fossil fuels	9,600
dissolved inorganic carbon in seawater	40,000
carbon in dead organic matter	
continental	1,500
marine	
deep ocean	2,000
mixed layer	40
CO_2 in the atmosphere (1980)	735
carbon in living organisms	
continental	560
marine	2

6. Major stocks of nitrogen

stock	magnitude [10^{12} kg(N)]
N_2 in the atmosphere	3.9×10^6
inorganic fixed nitrogen	
in soil	150
in seawater	350
in atmosphere	1.4
nitrogen in dead organic matter	
continental	100
marine	300
nitrogen in living organisms	
continental	7.5
marine	0.3

7. Major stocks of sulfur

stock	magnitude [10^{12} kg(S)]
dissolved SO_4^{-2} in seawater	1.4×10^6
sulfur in dead organic matter	50
sulfur in living organisms	3
sulfur in atmosphere (mostly SO_2, H_2S, SO_4^{-2})	0.004

8. Major stocks of phosphorus

stock	magnitude [10^{12} kg(P)]
phosphorus in living organisms	3
phosphorus in dead organic matter	25
inorganic phosphate in soil	200
dissolved and suspended phosphorus	
in the mixed ocean layer	3
in the deep ocean	100

XIV. Climate Data (see also Table VI.1)

1. Zonal climate parameters

latitude	surface air mean temperature (°C) January	July	average albedo	precipitation (m/yr)	area of zone (10^{12} m^2)	net flow of energy to zone (W/m^2)* Q_r	Q_e	Q_c	Q_o
80–90° N	-31	-1	0.65	0.19	3.9	-83	15	68	0
70–80° N	-25	2	0.5	0.26	11.5	-80	20	60	0
60–70° N	-22	12	0.41	0.97	18.8	-65	11	43	11
50–60° N	-10	14	0.39	0.72	25.5	-40	20	5	15
40–50° N	-1	20	0.35	0.78	31.6	-16	12	-5	9
30–40° N	11	26	0.32	0.77	36.7	5	-17	0	12
20–30° N	19	20	0.28	0.70	40.3	19	-41	21	1
10–20° N	25	28	0.25	1.17	42.8	31	-15	-3	-13
0–10° N	27	27	0.25	1.92	44.4	39	44	-51	-32
0–10° S	27	25	0.23	1.47	44.4	41	19	-32	-28
10–20° S	26	24	0.22	1.29	42.8	37	-21	-12	-4
20–30° S	25	18	0.25	0.85	40.3	27	-43	11	5
30–40° S	20	14	0.31	0.92	36.7	12	-25	5	8
40–50° S	12	8	0.34	1.02	31.6	-11	11	-8	8
50–60° S	5	1	0.41	0.97	25.5	-39	36	-12	15
60–70° S	0	-12	0.48	0.67	18.8	-81	40	41	0
70–80° S	-8	-30	0.57	0.25	11.5	-95	19	76	0
80–90° S	-13	-42	0.65	0.11	3.9	-95	8	87	0

*Q_r is the net radiation at the top of the atmosphere above the zone. A negative value means outgoing infrared plus reflected solar energy exceeds incoming solar energy. Q_e is the net latent heat flow to the zone, derived by subtracting annual evapotranspiration from precipitation and multiplying the difference by the latent heat of vaporiza-tion. Q_c is the net heat inflow to the zone convected poleward by atmospheric motions. Q_o is the net inflow of heat to the zone convected by ocean currents. A positive number represents an inflow to the zone; a negative number is an outflow. Within any zone, energy conservation leads to $Q_r + Q_e + Q_c + Q_o = 0$.

2. Albedos of selected surfaces on Earth

surface	albedo
snow	0.7 ± 0.2
sand	0.25 ± 0.05
grasslands	0.23 ± 0.03
bare soil	0.2 ± 0.05
forest	0.15 ± 0.1
water (highly dependent on surface roughness and incident angle of sunlight)	$0.2 + 0.6$ $- 0.2$

XV. Characteristics of "Standard" Adult Persons

characteristic	man	woman
mass (kg)	70	58
surface area (m^2)	1.8	1.6
total body water (% of mass)	60	50
total blood mass (kg)	5.5	4.1
breathing rate, resting (liters/min) (6–8 breaths/min)	7.5	6.0
breathing, light activity (liters/min) (12–14 breaths/min)	20	19
daily air intake (m^3)	22.8	21.1
daily water intake (kg)		
milk	0.30	0.20
tap water	0.15	0.10
other fluids	1.5	1.1
free water in food	0.70	0.45
from oxidation of food	0.35	0.25
protein intake (kg/day)	0.095	0.066
carbohydrate intake (kg/day)	0.39	0.27
fat intake (kg/day)	0.12	0.08
resting metabolic rate (J/sec-kg)	1.19	1.12
average food energy intake (J/day)	11.7×10^6	9.20×10^6
(or in terms of cal/day)	2.8×10^6	2.2×10^6
carbon dioxide exhaled (kg/day)	1.0	0.90

XVI. Human Population Estimates

year	global population (10^9 people)	U.S. population (10^9 people)
1650	0.5	
1850	1.1	0.023
1900	1.6	0.076
1910	1.7	0.092
1930	2.0	0.123
1950	2.5	0.152
1960	3.0	0.181
1970	3.6	0.205
1980	4.5	0.227
1983	4.7	0.234

Sources of the Data in the Appendix

References for each section are listed below, followed by the numbers of the subsections for which they were used. (Not all sections are subdivided.) Full references are listed in the Bibliography.

I. Weast (1982), 1–15.

II. Horvath (1970).
Weast (1982).

III. Clark, W. C. (1982).
Ehrlich et al. (1977).
Broecker (1974).

IV. Kittel et al. (1965).

V. Horvath (1970), 1.
National Oceanic and Atmospheric Administration (1976), 2.
Houghton (1977), 2.
Garrels et al. (1975), 3.
Fortescue (1980), 3.
Committee on the Atmosphere and the Biosphere (1981), 3.
Ehrlich et al. (1977), 3.

VI. Horvath (1970), 1.
Skinner (1969), 2 and 3.
Murray and Reeves (1977), 4.
Ehrlich et al. (1977), 4.

VII. Study of Man's Impact on Climate (1971), 1.
Houghton (1977), 1.
Ehrlich et al. (1977), 1.
United States Department of Commerce (1982), 1.
Hubbert (1969), 1 and 2.
Committee on the Atmosphere and the Biosphere (1981), 3.

VIII. Fortescue (1980), 1.
Hepler and Smith (1975), 2.
Lederer and Shirley (1978), 2.
Sargent-Welch, 2.

IX. Study of Man's Impact on Climate (1971).
Galloway et al. (1982).

X. Stumm and Morgan (1981), 1–5.
Weast (1982), 3.

XI. Lederer and Shirley (1978), 1.
Committee on Biological Effects of Ionizing Radiation (1980), 2.
United Nations Scientific Committee on the Effects of Atomic Radiation (1982), 2.

XII. Olson (1982), 1.
Whittaker (1970), 1 and 2.
Whittaker and Likens (1973), 1 and 2.
Ehrlich et al. (1977), 1 and 2.

XIII. Ehrlich et al. (1977), 1–8.
      Garrels et al. (1975), 1–8.
      Clark, W. C. (1982), 1 and 5.
      Committee on the Atmosphere and the Biosphere (1981) 2, 3, 5, and 6.

XIV. Budyko (1977), 1.
      Houghton (1977), 1.
      Study of Man's Impact on Climate (1971), 2.
      Neiburger et al. (1973), 2.

XV. International Commission on Radiological Protection (1975).

XVI. U.S. Department of Commerce (1984).

Glossary

activity the number of nuclear transformations per unit time of a collection of radioactive atoms.

adiabatic with no heat exchange. A system is said to undergo an adiabatic change if no heat flows in or out of the system. Adiabatic is also sometimes used to describe a change characterized by no sudden discontinuous behavior.

aerosol a suspension of finely divided liquids or solids in a gaseous medium.

age-specific death rates the per capita death rate of each age group in a population.

age-specific population the number of individuals within each age group in a population.

air shed a volume of air with boundaries chosen to facilitate determination of pollutant inflow and outflow. Boundaries are often chosen so that major sources of air pollution lie within the air shed.

albedo the fraction of light incident on a surface that is reflected by the surface.

alkalinity the capacity of negatively charged ions to neutralize acidity.

Allee effect the tendency of sparse populations of organisms to die out, as, for example, could occur because potential mates cannot locate each other or because the protection afforded by large herds is lacking.

anthropogenic resulting from human activities.

aphelion the point in a planetary orbit at which the planet is farthest from the Sun.

Avogadro's number 6.02×10^{23}, which is the number of molecules in a mole.

background concentration the concentration of a substance that would exist in the environment were it not for the presence of human beings.

basal metabolism the minimum energy consumption required by an organism at rest to survive.

bioconcentration the process by which certain chemicals become more concentrated in organisms at progressively higher levels in the food chain.

biomagnification (see bioconcentration)

blackbody a perfect absorber and radiator of electromagnetic radiation.

calorie a unit of energy: 1 calorie is the amount of heat needed to raise the temperature of 1 gram of water from 14.5°C to 15.5°C.

Calorie 10^3 calories.

Carnot engine an idealized engine capable of converting heat into work. It operates in a cycle consisting of repeated expansion (at high temperature) and compression (at low temperature) of a gas in a cylinder, at one end of which is a piston. The cycle can be reversed so that the engine functions as a refrigerator.

carrying capacity the maximum life-support capability of an ecosystem.

catalyst a substance that expedites a chemical reaction (or other process) but is not consumed in the process.

climate the weather patterns characteristic of a region.

codistill a substance is said to codistill if it evaporates along with its liquid medium.

consumption (of water) the use of water in such a way that it is rendered unavailable for direct further use.

convection the transfer of heat by the directed (nonrandom) movement of matter.

denitrification a microbial process that converts nitrate (NO_3) to molecular nitrogen (N_2) or nitrous oxide (N_2O).

density dependence a dependency of a per capita birth or death rate on the size of the population.

diffusion the gradual spreading of the molecules of one substance through some medium (often air or water). The spreading occurs in a direction from a region of high concentration of the substance toward a region of low concentration.

discount rate the rate at which money grows in value (relative to inflation) if it is invested.

donor-controlled flow a flow of a substance out of a compartment such that the rate of flow depends on the stock of the substance in the compartment. A linear donor-controlled flow rate is linearly proportional to the stock.

energy the capacity to produce heat.

entropy a measure of molecular disorder. This disorder can be manifested either as chaotic motion of molecules or as the mixing of various molecular species.

epilimnion the upper, warmer layer of a stratified lake.

equilibrium the state of a system in which there is no *net* inflow or outflow of matter or energy.

evapotranspiration the combination of evaporation and transpiration.

feedback a "circular" sequence of events in which a change in one system component triggers causes and effects (or stimuli and responses) throughout the system that in turn eventually alter the first component again. If the eventual change reinforces the original change the feedback is *positive*; if it tends to cancel the original change, the feedback is *negative*.

food chain a hierarchy of the organisms in an ecosystem organized according to who eats whom.

geothermal flux the energy flow, per unit area, from Earth's interior to the surface.

half-life the time it takes for an exponentially decreasing stock to diminish by a factor of one half.

heat the kinetic energy associated with the random movement of molecules.

heat of fusion the heat that must be extracted from liquid water at 0°C to convert it to ice at 0°C.

heat of vaporization the heat that must be added to liquid water at 100°C to convert it to vapor at 100°C.

Henry's constant the ratio of the aqueous concentration of a gas to the atmospheric concentration when the water and atmosphere are in equilibrium.

hydrocycle the flow of water on Earth among the major hydrologic compartments: oceans, groundwater, soil water, lakes, streams, ice, atmosphere, and organisms.

hypolimnion the lower, colder layer of a stratified lake.

incorporation efficiency the fraction of ingested food that is converted into body tissue and not excreted.

infrared radiation the range of electromagnetic radiation between visible radiation and microwaves (wavelengths between 0.78 microns and approximately 10^3 microns).

insolation the sunshine intensity on a given surface, usually measured in watts per square meter.

ions electrically charged atoms or molecules.

joule a unit of energy. A 1 kilogram object moving at a speed of 1 meter per second has a kinetic energy of 0.5 joules; also, 1 joule = 0.2390 calories.

kinetic energy for an object with mass m, moving at velocity v, the kinetic energy is $(mv^2)/2$ (unless the velocity approaches that of light, in which case the theory of relativity comes into play and this formula is modified).

lapse rate the rate of decrease of air temperature with increasing altitude.

latent heat the heat that is recoverable when water vapor condenses or ice melts.

Leslie matrix a matrix used to study the evolution through time of an age-specific population vector. The matrix multiplies the vector evaluated at one time to yield the value of the vector at a later time.

mass fraction the mass of an ingredient divided by the mass of the medium in which the ingredient occurs.

matrix an array (usually with an equal number of rows and columns) of numbers or other symbols.

mean-life the average lifetime of an entity in a collection of entities.

mobilization of trace substance the process by which a trace substance bound in soil, sediment, or rock enters water or the atmosphere. The mobilization of chemically or physically bound substances usually increases their biological accessibility.

mole (as used here, a gram-mole) M grams of a substance, where M is the substance's molecular mass. A mole contains Avogadro's number (6.02×10^{23}) of molecules.

nitrification a microbial process that converts NH_4^+ to NO_3^-.

optical coefficients the fractions of light incident on a material that are reflected, absorbed, and transmitted.

perihelion the point in a planetary orbit at which the planet is closest to the Sun.

photochemical reaction a chemical reaction triggered by sunlight.

photosynthate the organic product of photosynthesis.

primary productivity the rate at which new plant biomass is formed by photosynthesis. Gross primary productivity is the total rate of photosynthetic production of biomass; net primary productivity is gross primary productivity minus the respiration rate.

radiation the transfer of energy through space. A material medium is not necessary for this process.

receptor-controlled flow a flow of a substance into a compartment such that the rate of flow depends on the stock of the substance in the compartment. A linear receptor-controlled flow rate is linearly proportional to the stock.

residence time   the average time during which individual entities comprise a stock. In the steady state, residence time equals the size of the stock divided by the inflow or outflow rate.

respiration   the intake of oxygen and the production of carbon dioxide by organisms.

retention factor   a measure of the extent to which an organism retains in its body tissue a trace substance found in its food.

semipermeable membrane   a membrane through which water can pass but certain ions cannot.

significant figures   the digits in a numerical expression for a measured or estimated quantity that can be relied upon because of the accuracy and precision of the measurement or estimation.

steady state   a condition of balance, in which inflows equal outflows.

stoichiometric constants   the relative numbers of molecules of each chemical species that participate in a chemical reaction.

Stokes Law   this law states that the frictional force resisting the movement of an object in a medium such as air or water is proportional to the product of the average diameter of the object, the speed of the object, and the viscosity of the medium. The law is valid only if the object's speed is sufficiently slow.

stratosphere   the atmospheric layer above the troposphere. It extends from about 12 to 50 km altitude. Within it, temperature increases with altitude.

submicron   referring to particles of diameter less than 1 micron.

thermocline   the transition zone between the epilimnion and the hypolimnion of a lake or between the ocean's mixed surface layer and the deep layer.

tropopause   the boundary between the troposphere and the stratosphere. It lies at an altitude of about 8 km at the poles and 16 km at the equator.

troposphere   the lowest layer of the atmosphere, in which temperature decreases with altitude. It extends from Earth's surface to the temperature minimum at the tropopause.

transpiration   the passage of water to the atmosphere from the above-ground portion of green plants.

urban heat island effect   the increase in air temperature that results from intense fossil fuel consumption and from altered land surface characteristics in large urban areas.

vector   a linear array of numbers or other symbols.

Verhulst effect   a density-dependent death rate that depends quadratically on the size of the population.

viscosity   a measure of the "thickness" (or resistance to motion of an object moving through it) of a fluid such as air or water.

watershed   a region with boundaries chosen to facilitate determination of water inflow and outflow.

watt   a unit of power (or rate of energy flow) equal to 1 joule per second.

weather   the temperature, solar intensity, air pressure, precipitation, winds, humidity, and cloud conditions at a given time and place.

withdrawal (of water)   the removal of water from a water supply. The water may be consumed or returned to the water supply in a condition suitable for further use.

Bibliography

Ackerman, T. P. 1979. On the effect of CO_2 on atmospheric heating rates. *Tellus* 31:115–23.

Alvarez, L. W., Alvarez, W., Asaro, F., Michel, H. V. 1980. Extraterrestrial cause for the Cretaceous-Tertiary extinction. *Science* 208:1095–1108.

Augustsson, T., Ramanathan, V. 1977. A radiative-convective model study of the CO_2 climate problem. *J. Atmospher. Sci.* 34:448–51.

Bannister, J., Preston, S. H. 1981. Mortality in China. *Pop. Devel. Rev.* 7(1):98–110.

Birkhoff, G., MacLane, S. 1953. *A survey of modern algebra.* New York: Macmillan. pp. 112–13.

Bolin, B. 1974. Modeling the climate and its variations. *Ambio* 3(5):180–88.

Boyce, W. E., DiPrima, R. C. 1973. *Elementary differential equations and boundary value problems.* New York: Wiley.

Broecker, W. S. 1974. *Chemical oceanography.* New York: Harcourt Brace Jovanovich.

Budyko, M. I. 1977. *Climatic changes.* Washington, D.C.: Am. Geophys. Union.

Chamberlain, J. W. 1978. *Theory of planetary atmospheres: an introduction to their physics and chemistry.* New York: Academic Press.

Changnon, S. A. 1976. Inadvertent weather modification. *Water Resour. Bull.* 12(4):695–718.

Charlson, R. J., Rodhe, H. 1982. Factors controlling the acidity of natural rainwater. *Nature* 295:683–85.

Clark, C. 1977. The economics of whaling: a two-species model. In *New directions in the analysis of ecological systems,* ed. G. Innis. La Jolla, Calif.: Soc. Comput. Simulat.

Clark, W. C. 1982. *Carbon dioxide review 1982.* Oxford: Clarendon Press.

Committee on Biological Effects of Ionizing Radiations. 1980. *The effects on populations of exposure to low levels of ionizing radiation: 1980.* Washington, D.C.: Natl. Acad. Sci., Natl. Res. Counc., Natl. Acad. Press.

Committee on the Atmosphere and the Biosphere, Commission on Natural Resources. 1981. *Atmosphere-biosphere interactions: toward a better understanding of the ecological consequences of fossil fuel combustion.* Washington, D.C.: Natl. Acad. Sci., Natl. Res. Counc., Natl. Acad. Press.

Cronan, C., Schofield, C. L. 1979. Aluminum leaching response to acid precipitation: effects on high-elevation watersheds in the Northeast. *Science* 204:304–5.

Davidson, C., Grimon, T., Nasta, M. 1981. Airborne lead and other elements derived from local fires in the Himalayas. *Science* 214:1344–46.

Ehrlich, P. R., Ehrlich, A. H., Holdren, J. P. 1977. *Ecoscience: population, resources, environment*. San Francisco: W. H. Freeman.

Ehrlich, P. R., Harte, J., Harwell, M. R., Raven, P. H., Sagan, C., Woodwell, G. M., Berry, J., Ayensu, E. S., Ehrlich, A. H., Eisner, T., Gould, S. J., Grover, H. O., Herrera, R., May, R. M., Mayr, E., McKay, C. P., Mooney, H., Myers, N., Pimentel, D., Teal, J. M. 1983. Long-term biological consequences of nuclear war. *Science* 222:1293–1300.

Ehrlich, P. R., Ehrlich, A. H. 1981. *Extinction*. New York: Random House.

Eisenbud, M. 1963. *Environmental radioactivity*. New York: McGraw Hill.

Fortescue, J. A. C. 1980. *Environmental geochemistry: a holistic approach*. New York: Springer-Verlag.

Galloway, J., Thornton, J. D., Norton, S. A., Valchok, H., McLean, R. A. N. 1982. Trace metals in atmospheric deposition: a review and assessment. *Atmospher. Environ.* 16(7):1677–1700.

Garrels, R. M., Mackenzie, F. T., Hunt, C., 1975. *Chemical cycles and the global environment*. Los Altos, Calif.: William Kaufmann, Inc.

Goodman, N. 1956. *The ingenious Dr. Franklin: selected scientific letters of Benjamin Franklin*. Philadelphia: Univ. of Pennsylvania Press.

Goody, R. M., Walker, J. C. G. 1972. *Atmospheres*. Englewood Cliffs, N.J.: Prentice Hall.

Harte, J., Socolow, R. H. 1971. *Patient earth*. New York: Holt, Rinehart and Winston.

Harte, J., El-Gasseir, M. 1978. Energy and water. *Science* 199:623–34.

Hepler, L. G., Smith, W. L. 1975. *Principles of Chemistry*. New York: MacMillan.

Horvath, A. L. 1970. *Physical properties of inorganic compounds: SI units*. London: Edward Arnold.

Houghton, J. T. 1977. *The physics of atmospheres*. Cambridge: Cambridge Univ. Press.

Hubbert, M. K. 1969. Energy resources. In *Resources and man*. San Francisco: W. H. Freeman.

International Commission on Radiological Protection. 1975. *Reference man: anatomical, physiological, and metabolic characteristics*. ICRP Publ. No. 23. Oxford: Pergamon Press.

Keeling, C. D., Bacastow, R. B., Whorf, T. P. 1982. Measurements of the concentration of carbon dioxide at Mauna Loa Observatory, Hawaii. See Clark 1982.

Kerr, R. 1983a. Orbital variations—ice age link strengthened. *Science* 219:272–74.

Kerr, R. 1983b. Trace gases could cause climate warming. *Science* 220:1364–65.

Keyfitz, N. 1968. *Introduction to the mathematics of population*. Reading, Mass.: Addison-Wesley.

Keyfitz, N. 1984. The population of China. *Sci. Am.* 250:38–47.

Kittel, C., Knight, W. D., Ruderman, M. A. 1965. *Mechanics*. New York: McGraw Hill.

Lederer, C. M., Shirley, V. S. 1978. *Table of isotopes, 7th ed.* New York: Wiley.

Manabe, S., Wetherald, R. 1967. Thermal equilibrium of the atmosphere with a given distribution of relative humidity. *J. Atmosph. Sci.* 24:241–59.

May, R. M., Beddington, J. R., Clark, C. W., Holt, S. J., Laws, R. M. 1979. Management of multispecies fisheries. *Science* 205:267–77.

May, R. M. 1973. *Stability and complexity in model ecosystems.* Princeton, N.J.: Princeton Univ. Press.

Morowitz, H. 1970. *Entropy for biologists.* New York: Academic Press.

Murray, C. R., Reeves, E. B. 1977. *Estimated use of water in the United States in 1975.* U.S. Geol. Surv. Circ. 765. Washington, D.C.: U.S. Government Printing Office.

National Oceanic and Atmospheric Administration, 1976. *U.S. standard atmosphere.* Washington, D.C.: Natl. Aeronaut. Space Admin.

Neiberger, M., Edinger, J. D., Bonner, W. D. 1973. *Understanding our atmospheric environment.* San Francisco: Freeman.

Nero, A. V. 1983. Indoor air exposures from Rn^{222} and its daughters. *Health Phys.* 45(2):277–88.

Olson, J. 1982. Earth's vegetation and atmospheric carbon dioxide. See Clark 1982.

Olsson, M., Wick, G. L., Issass, J. P. 1979. Salinity gradient power: utilizing vapor pressure differences. *Science* 206:452–54.

Oppenheimer, M. 1983a. The relationship of sulfur emissions to sulfate in precipitation. *Atmospher. Environ.* 17:451–60.

Oppenheimer, M. 1983b. The relationship of sulfur emissions to sulfate in precipitation. II. Gas phase processes. *Atmospher. Environ.* 17:1489–95.

Pollack, J. B., Toon, O. B., Ackerman, T. P., McKay, C. P., Turco, R. P. 1983. Environmental effects of an impact-generated dust cloud: implications for the Cretaceous-Tertiary extinctions. *Science* 219:287–89.

Rasool, S. I., Schneider, S. H. 1971. Atmospheric carbon dioxide and aerosols: effects of large increases on global climate. *Science* 173:138–41.

Rohsenow, W. M., Hartnett, J. P. 1973. *Handbook of heat transfer.* New York: McGraw-Hill.

Rosen, H., Novakov, T., Bodhaine, B. 1981. Soot in the Arctic. *Atmospher. Environ.* 15:1371–74.

Ruddiman, W. F., McIntyre, A. 1981. Oceanic mechanism for amplification of the 23,000-year ice-volume cycle. *Science* 212:617–25.

Sagan, C., Toon, O. B., Pollack, J. B. 1979. Anthropogenic albedo changes and the earth's climate. *Science* 206:1363–68.

Sargent-Welch Scientific Co. 1980. *Table of periodic properties of the elements.* Catalog no. S-18806. Skokie, Ill.: Sargent-Welch.

Schmidt-Nielsen, K. 1972. *How animals work.* London: Cambridge Univ. Press.

Schneider, S. H. 1972. Cloudiness as a global climatic feedback mechanism: the effects on the radiation balance and surface temperature of variations in cloudiness. *J. Atmospher. Sci.* 29:1413–22.

Singh, H. B., Salas, L. J., Shigeishi, H., Scribner, E. 1979. Atmospheric halocarbons, hydrocarbons, and sulfur hexafluoride: global distributions, sources, and sinks. *Science* 203:899–903.

Skinner, B. J. 1969. *Earth Resources.* Englewood Cliffs, N.J.: Prentice Hall.

Strain, B., Armentano, T. V. 1980. Environmental and societal consequences of carbon dioxide induced climate change: response of 'unmanaged' ecosystems. Durham, N.C.: Duke Univ. Phytotron.

Study of Man's Impact on Climate. 1971. *Inadvertent climate modification.* Cambridge, Mass.: MIT Press.

Stumm, W., Morgan, J. J. 1981. *Aquatic chemistry,* 2nd ed. New York: Wiley.

Swartz, C. E. 1973. *Used math.* Englewood Cliffs, N.J.: Prentice Hall.

Turco, R. P., Toon, O. B., Ackerman, T. P., Pollack, J. B., Sagan, C. 1983. Nuclear winter: global consequences of multiple nuclear explosions. *Science* 222:1283–92.

United Nations Scientific Committee on the Effects of Atomic Radiation. 1982. *Ionizing radiation: sources and biological effects.* New York: United Nations.

United States Department of Commerce. 1982. *Statistical abstract of the United States*. Washington, D.C.: United States Government Printing Office.

Van Ness, H. C. 1969. *Understanding thermodynamics*. New York: McGraw-Hill.

Walker, J. C. G. 1977. *Evolution of the atmosphere*. New York: Macmillan.

Weast, R. C. 1982. *Handbook of chemistry and physics: a ready-reference book of chemical and physical data*. 63rd ed. Boca Raton, Fla.: CRC Press.

Wetzel, R. 1975. *Limnology*. Philadelphia: W. B. Saunders.

Woodwell, G. M., Hobbie, J. E., Houghton, R. A., Melillo, J. M., Moore, B., Peterson, B. J., Shaver, G. R. 1983. Global deforestation: contribution to atmospheric carbon dioxide. *Science* 222:1081–86.

Whittaker, R. H. 1970. *Communities and ecosystems*. New York: Macmillan.

Whittaker, R. H., Likens, G. 1973. Carbon in the biota. In *Carbon and the biosphere*, ed. G. M. Woodwell, E. V. Pecan. Springfield, Va.: Natl. Tech. Info. Serv., CONF-720510.

Zimmerman, P. R., Greenberg, J. P., Wardiga, S. O., Crutzen, P. J. 1982. Termites: a potentially large source of atmospheric methane, carbon dioxide, and molecular hydrogen. *Science* 218:563–65.

Answers To Exercises

I.
  1.2 10^3
  1.3 10^2
  1.4 10^{-4}

  2.1 5.06 cm^3
  2.2 3 Å

  3.1 500 m
  3.2 *(a)* 5×10^{-7}; *(b)* 6×10^{-5}
  3.3 3.94
  3.4 *(a)* $3.6 \times 10^{18} \text{ m}^2$; *(b)* $1.8 \times 10^{17} \text{ m}^2$

  4.1 gas: 170 yr; coal: 2800 yr
  4.2 15,000 yr

  5.1 1.5×10^{12}
  5.2 $\log_e 2 = 1/\log_2 e = 0.693$; $\log_{10} 2 = 1/\log_2 10 = 0.301$; $\log_e 10 = 1/\log_{10} e = 2.303$
  5.3 8 times
  5.4 0.405

  6.1 $f = 0.0056$
  6.3 0.1
  6.4 2236

  7.1 Cd: 300 tonnes; Pb: 30,000 tonnes; Zn: 150,000 tonnes; Se: 3,000 tonnes; Hg: 300 tonnes; As: 30,000 tonnes

II.
  1.1 42 cows
  1.2 *(a)* 4 yr; *(b)* 4.75 yr

  2.1 2960 yr
  2.2 40 in

  3.1 1.3
  3.3 *(a)* 1/14; *(b)* 14/100

  4.2 $2.1 \times 10^{11} \text{ kg}(N_2O)/\text{yr}$

  6.2 *(a)* $S_A - E_A$; $S_A - E_A + S_B$; *(b)* V_A/S_A; $V_B/(S_A - E_A + S_B)$; *(c)* $C_A = P/(S_A - E_A)$ g/liter; $C_B = P/(S_A + S_B - E_A - E_B)$ g/liter

7.1 1.2×10^6 tonnes/yr
7.2 $\alpha^{-1} = 1.125$ yr
7.3 $\beta^{-1} = 1.125$ yr
7.4 $\alpha^{-1} = 2.8$ yr
7.5   1 or 2%
7.6   (a) 0.6×10^{11} moles/yr; (b) 0.8×10^{11} moles/yr

8.2 $\overline{X}_1' = 0.183,$
          $\overline{X}_2' = 0.1, \overline{X}_3' = 1.017$
8.3 $\overline{X}_1' = 0.8\,\overline{X}_1; \overline{X}_2' = \overline{X}_2; \overline{X}_3' = 1.08\,\overline{X}_3$

9.1   2.4% decrease

10.2   (a) 53 kg(fuel)/km^{2}day; (b) order of magnitude of 10 people/km^2

11.3   (a) 7×10^{-10} Ci/m^3; (b) 9.4×10^{-8} Ci/m^3
11.4 6.4×10^7 atoms
11.5 1.1×10^{-14} ppm(v)

12.1   0.001
12.2   4%

13.2   10.9 W/m^2 (180 × Earth's)
13.3 1.16×10^9m

14.1   (a) 100 W; (b) 75 W
14.2   250 W

15.1   0.0083

16.1   15%
16.2   160 km^2
16.3   5/3
16.5   180 km

17.1 $n = 1.1 \times 10^3$ moles/m^3, $P = 2.65 \times 10^6$ J/m^3
17.2   (a) 2/3; (b) 1300 yr

18.2   0.713
18.3 $f_S = (1 - R')[T/(1 - R'R)]$, $f_C = (1 - R - T)[1 + R'T/(1 - R'R)]$
18.5   high particles: $a = T^2R' + (TT')^2R''/(1 - R'R'')$, $f_S = TT'A''/(1 - R'R'')$
           low particles: $a = R' + (TT')^2R''/(1 - T^2R'R'')$, $f_S = TT'A''/(1 - T^2R'R'')$

19.2   0.78%
19.3   0.0017%

20.1   a decrease of 0.046 pH units
20.2   (a) $x^3 \approx 125$; (b) dropping the x^3 and 10x terms, x = $(4/3)^{1/2}$; dropping only
           the x^3 term, x = 4/3
20.4   (a) 5.1×10^{-8} g; (b) pH = 8.26

21.1   (a) 2.4×10^{14} moles(H$^+$)/yr; (b) pH = 2.66

22.1   coal: 120 yr; gas: 98 yr

23.1   400 tonnes; 400 ppm by weight
23.4   1.4 m

III.   **1.2**   3.82
           1.3   5.6

           2.2   (liters/mole)^2
           2.3   2.1 ppm by weight
           2.4 1.4×10^6 g/yr
           2.5 2.4×10^4 yr

           3.3   1563 yr

3.4 8220 yr
3.5 9 yr
3.6 8.3×10^9 tonnes/yr
3.7 (a) 8.6 years; (b) about 7.4 yr

4.2 $dY/dX = 2.3 \times 10^{-21} M_s$
4.3 -0.14 pH units
4.5 957 micromoles/liter

6.2 $g/C_p = 9.76$ K/km
6.4 $T_0 = 255$ K, $T_1 = 253$ K, $T_s = 243$ K
6.3 $T_0 = T_1 = T_s = T_{dust} = 221.8$ K
6.5 (c) 81.0%; 19.6%; 5.47 km
      (d) 80.5%; 17.2%; 5.67 km
6.6 roughly 1.7 km

7.3 $T_0 = 249.6$ K, $T_1 = 278.5$ K, $T_s = 289.7$ K
7.4 70% increase

8.2 $(1 - A)^{-1} = 0.61$
8.6 $(1 - A)^{-1} = 3.7$

9.2 345 km $\times$ 345 km

13.3 $\bar{n} = r/(1 - r)$
13.4 first method: $T = 200$ days; second method: $T = 250$ days

14.5 (a) 74%; (b) 54%; (c) 20%
14.7 $X = 0.525884288$

Index